Photovoltaic Functional Crystals and Ceramics

Photovoltaic Functional Crystals and Ceramics

Guest Editors

**Linghang Wang
Gang Xu**

Basel • Beijing • Wuhan • Barcelona • Belgrade • Novi Sad • Cluj • Manchester

Guest Editors

Linghang Wang
Xi'an Jiaotong University
Xi'an
China

Gang Xu
Xi'an Technological University
Xi'an
China

Editorial Office
MDPI AG
Grosspeteranlage 5
4052 Basel, Switzerland

This is a reprint of the Special Issue, published open access by the journal *Crystals* (ISSN 2073-4352), freely accessible at: https://www.mdpi.com/journal/crystals/special_issues/photovoltaic_ceramics.

For citation purposes, cite each article independently as indicated on the article page online and as indicated below:

Lastname, A.A.; Lastname, B.B. Article Title. *Journal Name* **Year**, *Volume Number*, Page Range.

ISBN 978-3-7258-2833-3 (Hbk)
ISBN 978-3-7258-2834-0 (PDF)
https://doi.org/10.3390/books978-3-7258-2834-0

© 2024 by the authors. Articles in this book are Open Access and distributed under the Creative Commons Attribution (CC BY) license. The book as a whole is distributed by MDPI under the terms and conditions of the Creative Commons Attribution-NonCommercial-NoDerivs (CC BY-NC-ND) license (https://creativecommons.org/licenses/by-nc-nd/4.0/).

Contents

Linghang Wang and Gang Xu
Progress in the Applications of Photovoltaic Functional Crystals and Ceramics
Reprinted from: *Crystals* 2024, 14, 958, https://doi.org/10.3390/cryst14110958 1

Bin Yu, Chenggang Xu, Mingxing Xie, Meng Cao, Jijun Zhang, Yucheng Jiang and Linjun Wang
Deposition of CdZnTe Films with CSS Method on Different Substrates for Nuclear Radiation Detector Applications
Reprinted from: *Crystals* 2022, 12, 187, https://doi.org/10.3390/cryst12020187 5

Gang Xu, Ming Yao, Mingtao Zhang, Jinmeng Zhu, Yongxing Wei, Zhi Gu and Lan Zhang
Evaluation of Relationship between Grain Morphology and Growth Temperature of HgI_2 Poly-Films for Direct Conversion X-ray Imaging Detectors
Reprinted from: *Crystals* 2022, 12, 32, https://doi.org/10.3390/cryst12010032 16

Maxim V. Zdorovets, Dmitriy I. Shlimas, Artem L. Kozlovskiy and Daryn B. Borgekov
Effect of Irradiation with Low-Energy He^{2+} Ions on Degradation of Structural, Strength and Heat-Conducting Properties of BeO Ceramics
Reprinted from: *Crystals* 2022, 12, 69, https://doi.org/10.3390/cryst12010069 24

Ling Gao, Ruidong Guan, Shengnan Zhang, Hao Zhi, Changqing Jin, Lihua Jin, et al.
As-Sintered Manganese-Stabilized Zirconia Ceramics with Excellent Electrical Conductivity
Reprinted from: *Crystals* 2022, 12, 620, https://doi.org/10.3390/cryst12050620 36

Cuixia Liu and Yu Yao
Study of Crack-Propagation Mechanism of $Al_{0.1}CoCrFeNi$ High-Entropy Alloy by Molecular Dynamics Method
Reprinted from: *Crystals* 2023, 13, 11, https://doi.org/10.3390/cryst13010011 44

Cuixia Liu, Rui Wang and Zengyun Jian
The Influence of Grain Boundaries on Crystal Structure and Tensile Mechanical Properties of $Al_{0.1}CoCrFeNi$ High-Entropy Alloys Studied by Molecular Dynamics Method
Reprinted from: *Crystals* 2022, 12, 48, https://doi.org/10.3390/cryst12010048 54

Yingxue Xi, Jiwu Zhao, Jin Zhang, Changming Zhang and Qi Wu
The Influence of Ion Beam Bombardment on the Properties of High Laser-Induced Damage Threshold HfO_2 Thin Films
Reprinted from: *Crystals* 2022, 12, 117, https://doi.org/10.3390/cryst12010117 66

Jongbeom Lee, Jinyoung Jeong, Hyowon Lee, Jaesoung Park, Jinman Jang and Haguk Jeong
Effects of Sintering Processes on Microstructure Evolution, Crystallite, and Grain Growth of MoO_2 Powder
Reprinted from: *Crystals* 2023, 13, 1311, https://doi.org/10.3390/cryst13091311 74

Daryn B. Borgekov, Artem L. Kozlovskiy, Rafael I. Shakirzyanov, Ainash T. Zhumazhanova, Maxim V. Zdorovets and Dmitriy I. Shlimas
Properties of Perovskite-like Lanthanum Strontium Ferrite Ceramics with Variation in Lanthanum Concentration
Reprinted from: *Crystals* 2022, 12, 1792, https://doi.org/10.3390/cryst12121792 85

Mohammed Sobhy and F. H. H. Al Mukahal
Analysis of Electromagnetic Effects on Vibration of Functionally Graded GPLs Reinforced Piezoelectromagnetic Plates on an Elastic Substrate
Reprinted from: *Crystals* **2022**, *12*, 487, https://doi.org/10.3390/cryst12040487 . . . . **98**

Editorial

Progress in the Applications of Photovoltaic Functional Crystals and Ceramics

Linghang Wang [1,*] and Gang Xu [2]

1. Electronic Materials Research Laboratory, Key Laboratory of the Ministry of Education and International Center for Dielectric Research, School of Electronic Science and Engineering, Xi'an Jiaotong University, Xi'an 710049, China
2. School of Materials and Chemical Engineering, Xi'an Technological University, Xi'an 710021, China; xxrshhuangshan@126.com
* Correspondence: lhwang@xjtu.edu.cn

Citation: Wang, L.; Xu, G. Progress in the Applications of Photovoltaic Functional Crystals and Ceramics. *Crystals* 2024, 14, 958. https://doi.org/10.3390/cryst14110958

Received: 24 October 2024
Accepted: 30 October 2024
Published: 1 November 2024

Copyright: © 2024 by the authors. Licensee MDPI, Basel, Switzerland. This article is an open access article distributed under the terms and conditions of the Creative Commons Attribution (CC BY) license (https://creativecommons.org/licenses/by/4.0/).

1. Introduction

With the progression of mankind and the development of technology, great strides have been made regarding the application of inorganic crystalline materials in a number of fields such as high-energy and nuclear physics, environmental and safety inspection, the optoelectronics and communication fields, energy, and aerospace engineering, particularly the industrialization of photovoltaic and detector materials, which has brought mankind's knowledge of natural disciplines to an all-time high. This has further promoted interdisciplinary collaboration between the fields of optoelectronic functional devices, biomedical engineering, and materials science, and has prompted scholars to accelerate the in-depth study of the original mechanisms of the physical responses of these materials in these fields and the commercialization of such functional devices.

This Editorial refers to the Special Issue "Inorganic Crystalline Materials". This Special Issue aims to highlight new opportunities and challenges for advancing the development of the Inorganic Crystalline Materials, focusing on crystal or film growth, characterization, structure refinement, modeling, device fabrication and measurements, and system testing, as well as corresponding research which is fundamental to the field.

After repeated correspondence between the reviewers and authors to ensure the reference value of the papers, a total of 10 articles were finally accepted. This provides a platform for academic exchange and keeps us abreast of the research results of our peers.

2. An Overview of the Published Articles

As we all know, medical imaging detection technology has developed rapidly in recent years. A number of detectors, including cardiac-specific SPECT systems, bone densitometers, CT imaging, and soft-tissue imaging techniques, etc., have been developed for the detection of Cadmium Zinc Telluride (CZT), all of which have increased the efficiency of detecting this compound exponentially, with improved system sensitivity and image quality, and have greatly improved the accuracy of diagnosis. Bin Yu et al. [1] investigated the growth characteristics of CZT thin films on different substrates using the near-space sublimation method. They found that films grown on a (111)-oriented CZT wafer show good reactions to nuclear radiation signals and can more effectively detect radiation from weak radiation sources compared to films grown on non-oriented CZT wafers and FTO substrates. This result suggests that substrate selection plays a crucial role in the development of thin-film devices in terms of film quality and even device performance. Mercury iodide crystals, a crystal material used for medical imaging, also have a long history in crystals research. Compared with CZT, mercury iodide crystals have advantages in terms of energy resolution, particularly for applications in mammography and digital X-ray imaging. This is likely the main reason why researchers have been focusing on mercury iodide crystals.

Gang Xu et al. [2] reported the preparation of large-area mercury iodide thin-film imaging detectors and discussed the relationship between the deposition temperature and the quality of the film, creating a positive reference for further optimizing this process to obtain large-area thin-film devices for medical imaging and for the commercial development of such devices.

BeO ceramics are one of the most popular choices of structural materials in the nuclear energy industry due to their high thermal conductivity, high strength, high insulation capacity, chemical stability, and high temperature resistance. The volume expansion of BeO ceramics and the microcrack generation mechanism, as well as the design of ceramic structures, under irradiation conditions have been hot topics in research into structural materials suitable for nuclear energy generation. Maxim V. Zdorovets et al. [3] investigated the radiation-induced damage kinetics of beryllium oxide ceramics under low-energy helium ion irradiation. The irradiation-induced structural changes were found to be related to the amorphization process and the increase in the dislocation density of the ceramics. In addition, a decrease in the hardness and wear resistance of the ceramics was also found. These results undoubtedly have a positive significance for the in-depth study of the compositional and structural design of ceramics under irradiation conditions.

Anisotropic-shaped zirconia structures such as nano-rods, nano-belts, or platelets are thought to be useful starting materials for the oriented growth of zirconia ceramics and the fabrication of shape-dependent zirconia catalysts or catalytic supports, luminescent materials, gate dielectrics, and solid-state oxide fuel cells. Anisotropic ZrO_2 particles with octahedron-, diamond-, and plate-like morphologies were successfully synthesized through a facile hydrothermal treatment approach using NaBF4 as mineralizer by Ling Gao et al. [4]. The results showed that F- plays an essential role in tuning the crystallinity and size of primary ZrO_2 nanorods along the [001] direction. The secondary particles mainly crystallize on the small primary nanoparticles through the oriented attachment mechanism. In addition, octahedron-like ZrO_2 particles have the highest MB degradation rate. Clearly, these results hold positive significance for exploring the potential utilization of these structures as photocatalytic materials and starting materials for preparing oriented polycrystalline zirconia ceramics.

Al0.1CoCrFeNi high-entropy alloys have excellent mechanical properties which are superior to those of traditional alloys. However, finding a suitable strengthening mechanism is still challenging. The tensile properties of high-entropy alloys and the crack propagation mechanism have been investigated using the molecular dynamics method by Cuixia Liu et al. [5,6]. They thought that, during the plastic deformation of high-entropy alloys, each dislocation nucleates and emission continues near the crack tip, alongside the formation of complementary stacking layers or twins. Therefore, they argue that atomic shear behavior is caused by a dislocation motion. The above viewpoints provide a theoretical basis for improving the mechanical properties of high-entropy alloys.

Hafnium dioxide (HfO_2) boasts excellent optical, thermal, and mechanical properties and is one of the most important high-refractive-index oxide materials used for manufacturing interference multilayer films, and is also known to be a material with a high laser damage threshold (LIDT). Electron beam physical vapor deposition (EB-PVD) is considered to be one of the most critical techniques for preparing multilayer interference films; however, films prepared using this method tend to be porous, and segregation may occur during fabrication, leading to a decrease in the HfO_2 film's resistance to laser damage. Yingxue Xi and colleagues [7] studied a series of ion beam processing methods, comparing the impact of argon ions and oxygen ions at different energies on the optical performance, laser damage resistance, and surface quality of the films. The study found that oxygen ions at certain energies can effectively increase the film's laser damage threshold, primarily because oxygen ions can enhance the film's density and adjust the composition of segregated elements within the film. This finding is of significant value for enriching the preparation of laser film thresholds and clarifying the factors associated with film thresholds.

Molybdenum oxides, a type of metal oxide with an n-type semiconducting and non-toxic nature, have attracted much attention due to their diverse functional applications in electronics, catalysis, sensors, energy-storage units, field emission devices, superconductor lubricants, thermal materials, biosystems, and chromogenic and electrochromic systems. A corresponding study about the sintering of MoO_2 micropowders was carried out by Jongbeom Lee et al. [8] The authors discussed the morphological transformation and grain size changes in MoO_2 micropowders under the sintering process, analyzed the corresponding physical mechanisms, and drew clear conclusions. This serves as a good reference, providing an in-depth explanation of the basic knowledge in this field. Another fundamental study by Daryn B. Borgekov [9] focuses on the field of alternative energy. He investigated the effect of a change in the lanthanum concentration (La) during the synthesis of perovskite-like ceramics based on strontium ferrite on phase formation and subsequent changes in the conductive and thermophysical parameters. The results show that impedance spectroscopy shows a significant increase in the permittivity of the synthetic ceramics as the concentration of lanthanum in the synthetic ceramics increases over a limited range; there is also an increase in the tangent of the dielectric loss at a defined frequency. Furthermore, the resistivity decreases by one order of magnitude.

Mohammed Sobhy [10] reported the design of a fundamental application of this technology. He proposed using piezoelectric and piezomagnetic materials for electromechanical energy and magnetic energy interconversion which can be applied in nanoelectromechanical systems such as heat exchangers, smart devices, nuclear devices, and nanogenerators. Based on a refined four-unknown shear deformation plate theory, the free vibration of piezoelectromagnetic plates reinforced with functionally graded graphene nanosheets (FG-GNSs) under simply supported conditions is analyzed. After a rigorous mathematical derivation and analysis, it was found that the elastic foundation stiffness, graphene weight fraction, applied magnetic potential, and electromagnetic properties of graphene enhance the plate stiffness, leading to a noticeable increment in the fundamental frequency. Conversely, the increase in the side-to-thickness ratio, plate aspect ratio, and applied electric potential weakened the plate strength; therefore, the frequency decreases. Such research can have positive value for continuing to optimize the system design subsequently.

3. Conclusions

This Special Issue contains a number of studies related to basic material design, semiconductor material applications, and the application and conversion of energy, as well as designs for functional devices, showing a wealth of research results and providing a platform for researchers to communicate with each other. As Guest Editors, we would like to thank all the authors for their contributions and efforts.

Funding: This research received no external funding.

Acknowledgments: As Guest Editor of the Special Issue "Inorganic Crystalline Materials". I would like to express my deep appreciation to all authors whose valuable work was published under this issue and thus contributed to the success of the edition.

Conflicts of Interest: The authors declare no conflicts of interest.

References

1. Yu, B.; Xu, C.; Xie, M.; Cao, M.; Zhang, J.; Jiang, Y.; Wang, L. Deposition of CdZnTe Films with CSS Method on Different Substrates for Nuclear Radiation Detector Applications. *Crystals* **2022**, *12*, 187. [CrossRef]
2. Xu, G.; Yao, M.; Zhang, M.; Zhu, J.; Wei, Y.; Gu, Z.; Zhang, L. Evaluation of Relationship between Grain Morphology and Growth Temperature of HgI_2 Poly-Films for Direct Conversion X-ray Imaging Detectors. *Crystals* **2022**, *12*, 32. [CrossRef]
3. Zdorovets, M.V.; Shlimas, D.I.; Kozlovskiy, A.L.; Borgekov, D.B. Effect of Irradiation with Low-Energy He^{2+} Ions on Degradation of Structural, Strength and Heat-Conducting Properties of BeO Ceramics. *Crystals* **2022**, *12*, 69. [CrossRef]
4. Gao, L.; Guan, R.; Zhang, S.; Zhi, H.; Jin, C.; Jin, L.; Wei, Y.; Wang, J. As-Sintered Manganese-Stabilized Zirconia Ceramics with Excellent Electrical Conductivity. *Crystals* **2022**, *12*, 620. [CrossRef]
5. Liu, C.; Yao, Y. Study of Crack-Propagation Mechanism of $Al_{0.1}CoCrFeNi$ High-Entropy Alloy by Molecular Dynamics Method. *Crystals* **2023**, *13*, 11. [CrossRef]

6. Liu, C.; Wang, R.; Jian, Z. The Influence of Grain Boundaries on Crystal Structure and Tensile Mechanical Properties of $Al_{0.1}CoCrFeNi$ High-Entropy Alloys Studied by Molecular Dynamics Method. *Crystals* **2022**, *12*, 48. [CrossRef]
7. Xi, Y.; Zhao, J.; Zhang, J.; Zhang, C.; Wu, Q. The Influence of Ion Beam Bombardment on the Properties of High Laser-Induced Damage Threshold HfO_2 Thin Films. *Crystals* **2022**, *12*, 117. [CrossRef]
8. Lee, J.; Jeong, J.; Lee, H.; Park, J.; Jang, J.; Jeong, H. Effects of Sintering Processes on Microstructure Evolution, Crystallite, and Grain Growth of MoO_2 Powder. *Crystals* **2023**, *13*, 1311. [CrossRef]
9. Borgekov, D.B.; Kozlovskiy, A.L.; Shakirzyanov, R.I.; Zhumazhanova, A.T.; Zdorovets, M.V.; Shlimas, D.I. Properties of Perovskite-like Lanthanum Strontium Ferrite Ceramics with Variation in Lanthanum Concentration. *Crystals* **2022**, *12*, 1792. [CrossRef]
10. Sobhy, M.; Al Mukahal, F.H.H. Analysis of Electromagnetic Effects on Vibration of Functionally Graded GPLs Reinforced Piezoelectromagnetic Plates on an Elastic Substrate. *Crystals* **2022**, *12*, 487. [CrossRef]

Disclaimer/Publisher's Note: The statements, opinions and data contained in all publications are solely those of the individual author(s) and contributor(s) and not of MDPI and/or the editor(s). MDPI and/or the editor(s) disclaim responsibility for any injury to people or property resulting from any ideas, methods, instructions or products referred to in the content.

Article

Deposition of CdZnTe Films with CSS Method on Different Substrates for Nuclear Radiation Detector Applications

Bin Yu [1,†], Chenggang Xu [2,†], Mingxing Xie [3], Meng Cao [2,4,*], Jijun Zhang [2,*], Yucheng Jiang [5] and Linjun Wang [2,6]

1. State Key Laboratory of Nuclear Power Safety Monitoring Technology and Equipment, China Nuclear Power Engineering Co., Ltd., Shenzhen 518124, China; yubin@cgnpc.com.cn
2. School of Materials Science and Engineering, Shanghai University, Shanghai 200072, China; chenggangxu25@163.com (C.X.); ljwang@shu.edu.cn (L.W.)
3. School of Humanity, Shanghai University of Finance and Economics, Shanghai 200433, China; songlun2088@163.com
4. Key Laboratory of Infrared Imaging Materials and Detectors, Shanghai Institute of Technical Physics, Chinese Academy of Sciences, Shanghai 200083, China
5. Jiangsu Key Laboratory of Micro and Nano Heat Fluid Flow Technology and Energy Application, School of Physical Science and Technology, Suzhou University of Science and Technology, Suzhou 215009, China; jyc@usts.edu.cn
6. Zhejiang Institute of Advanced Materials, SHU, Jiashan 314113, China
* Correspondence: caomeng@shu.edu.cn (M.C.); zhangjijun81@shu.edu.cn (J.Z.)
† These authors contributed equally to this work.

Abstract: CdZnTe (CZT) films were grown by closed space sublimation (CSS) method on (111)-oriented CZT wafers, non-oriented CZT wafers and FTO substrates. The compositional and morphological properties of CZT films on different substrates were characterized by scanning electron microscopy (SEM) and atomic force microscopy (AFM), which indicated that CZT films grown on (111)-oriented CZT wafers had low dislocation density and high Zn composition. X-ray diffraction (XRD) measurements confirmed that CZT films grown on (111)-oriented CZT wafers had the best crystal quality. The I-V and DC photoconductivity measurements indicated that CZT films on (111)-oriented CZT wafer had good carrier transport performance. The energy spectra of CZT films grown on (111)-oriented CZT wafer presented that it had a good response to the nuclear radiation under ^{241}Am.

Keywords: CdZnTe; single crystal; thin film; closed space sublimation

1. Introduction

Cadmium zinc telluride (CZT) is an II-VI compound semiconductor with excellent performance, which has attracted wide attention in the fabrication of detectors, especially on nuclear detection devices [1–4]. Due to its high atomic number (48, 30, 52), high resistivity (>10^{10} Ω cm), large bandgap (~1.57 eV) and high density (5.78 g/cm^3), CZT has been demonstrated to be the most promising room-temperature nuclear detector. Compared with Si and high-purity Ge detectors with low atomic number and low detection temperature (77 K), CZT detectors have advantages in detection efficiency and room temperature [5]. Compared with HgI$_2$ and CdTe, CZT detector is more stable and has no polarization effect [6,7]. So far, the research on CZT has mainly focused on the detector-grade CZT single crystal [8]. However, the single-crystal CZT bulk with a large-diameter, high μτ, and good homogeneity is difficult to achieve. Therefore, the cost of CZT crystal is high and the growth process is complex. Due to the similarity between CZT films and CZT single crystal in some applications, it is proposed to replace CZT single crystal with CZT thick films. For example, the electrical properties of CZT films are similar to those of CZT single crystal, but the growth process of CZT films is easier and the cost is lower. Therefore,

CZT thick films are also suitable for some important applications, such as nuclear medical imaging and environmental protection [9–11].

Usually, CZT films can be deposited with a physical vapor deposition method, such as vacuum evaporation method [12], metal organic chemical vapor deposition (MOCVD) [13], closed space sublimation (CSS) [14,15], liquid phase epitaxy (LPE) [16], and so on. Among these, the vacuum evaporation method needs a very high vacuum atmosphere, which is difficult to achieve. MOCVD requires a high purity source, which is relatively expensive Compared with the above methods, the CSS method has the advantages of low cost, fast growth rate, simple process and good film quality [17–19].

Some researchers have reported deposition of CZT films on GaAs substrate for radiation detectors [20]. However, GaAs substrate is expensive and difficult to deposit. Thus, the effect of substrates to the properties of CZT films still needs further study. In this paper, three kinds of substrate materials were selected for the growth of CZT films, namely (111)-oriented CZT wafer, non-oriented CZT wafer and fluorine-doped tin oxide (FTO). The surface morphology, structural and electrical properties of CZT films grown on different substrates were studied in detail, which contributes to their applications on the nuclear radiation detectors.

2. Experimental

CZT films were grown on (111)-oriented CZT wafers, non-oriented CZT wafers and FTO substrates by CSS method. FTO substrates should be cleaned with emulsifier and ultrasonic cleaned in acetone to ensure the cleanness of glass surface. CZT wafers were firstly polished (PM6, Logitech, Glasgow, UK) using alumina slurry with particle sizes of 0.3 μm and 0.05 μm for 10 min, respectively. Then, CZT wafers were chemically polished by using Br_2 + lactic acid + ethylene glycol for 2 min. Finally, CZT films were grown on the prepared substrate by CSS method (SK-CSS-400) with the vacuum of 10^{-3} Pa. High-purity CZT polycrystalline powder was used as sublimation source (>99.99999%), which was obtained by grinding the CZT polycrystalline powder grown by vertical Bridgman method. The distance between the source and the substrate was kept at 5 mm. The source temperature and the substrate temperature were 650 °C, and 400 °C, respectively. The deposition time was 3 h. The average thicknesses of the three grown CZT films were about 12~15 μm. After deposition, the properties of CZT films on different substrates were characterized. The structural properties of CZT films on different kinds of substrates were measured by X-ray diffraction (XRD, Rigaku Ultima IV). The crystal sizes of CZT films were calculated by Scherrer equation [21]:

$$L = \frac{K\lambda}{\beta \cos \theta} \quad (1)$$

where λ is wavelength of the X-ray source (0.154 nm), K is constant (0.94), θ is the diffraction angle, and β is the Full width at half-maximum of the peak., The micro stress of the film was calculated according to the following formula [22]:

$$\varepsilon = \frac{1}{4} W \cos \theta \quad (2)$$

where W is the FWHM of CZT (111) surface. The morphologies of CZT films were measured by scanning electron microscopy (SEM, JSM-7500f) and Atomic Force Microscopy (AFM, Leap 4000x HR) and transmission electron microscope (TEM, JEM2100). The atomic compositions of CZT films were determined by X-ray photoelectron spectroscopy (XPS, ESCALAB 250Xi, Thermo SCIENTIFIC, Waltham, MA, USA). During the XPS measurement, the excitation source was Al Kα. The energy of X-ray was 1486.6 eV and X-ray tube voltage was 15 kV.

The Au contacts on CZT films were deposited with the electroless $AuCl_3$ technique. An $AuCl_3$ solution with an $AuCl_3$ to dioxide H_2O ratio of 1:25 was used for the electroless deposition. [23] The square Au contacts with area of 8 × 8 mm² were deposited on both sides of the CZT films. The electrical properties of CZT films were measured by Keithley 2400. The planar detectors of CdZnTe films were fabricated for energy spectroscopy response measurements and were irradiated by ^{241}Am sources (59.5 keV) at room temperature. The detectors were connected to an ORTEC 142IH preamplifier, and then the energy spectra of the CZT films were revealed by an oscilloscope and a Trump-PCI-2K multichannel analyzer system. The bias voltage applied to CZT films detectors is 50 V, the shaping time is 2 μs, and the amplification factor through preamplifier is 250 times.

3. Results and Discussions

3.1. XRD Patterns of CZT Films on Different Substrates

Figure 1 shows the XRD patterns of CZT films grown on (111)-oriented CZT, non-oriented CZT and FTO substrates. The diffraction peak intensity at 2 theta ≈ 24° is the highest, which corresponds to (111) plane of CZT. It indicates that the preferred orientation of the films is (111) face. The CZT films grown on (111)-oriented CZT, non-oriented CZT and FTO substrates show diffraction peaks at 2 theta = 24.16°, 23.64° and 23.60°, respectively [24]. The FWHMs of (111) face CZT on (111)-oriented CZT, non-oriented CZT and FTO substrates are 0.142°, 0.171° and 0.202°. The diffraction peak of CZT films on (111)-oriented CZT is also the strongest. Therefore, the quality of the films grown on the (111)-oriented CZT wafer is the best. CZT films grown on the non-oriented CZT wafer also have good quality, but the diffraction peak is relatively weak. For CZT films deposited on non-oriented CZT wafer, there is a weak peak at 2 theta = 56.81° and its FWHM is 0.289°. It may be the (400) crystal plane of CdTe, which is related to the crystal orientation of the non-oriented CZT wafer.

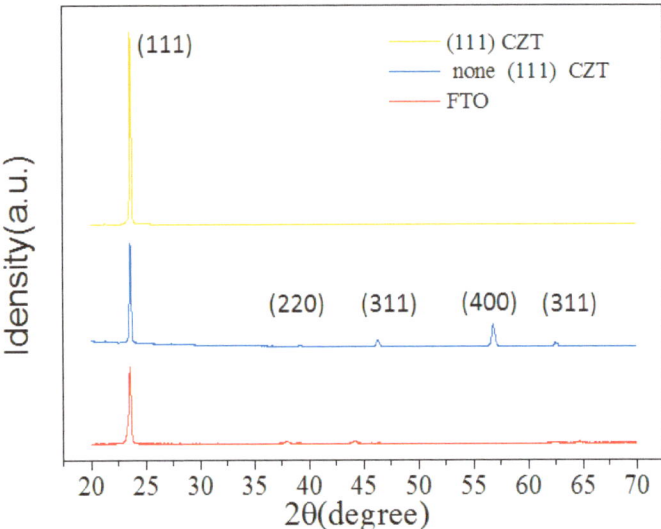

Figure 1. XRD patterns of CZT films deposited on different substrates.

The crystal sizes of CZT films can be calculated by Scherrer equation according to Equation (1). The crystal sizes of CZT films grown on (111)-oriented CZT wafer, non-oriented CZT wafer and FTO substrates are determined to be 72.4 nm, 60.1 nm and 50.8 nm, respectively. The micro stress in the films will enlarge the diffraction peak of the material, and some micro stress will cause additional displacement of the diffraction peak value of 2 theta, which will affect the accuracy of the measurement of the residual stress. According to the Equation (2), the micro stress of CZT film was calculated. The micro stress of CZT films grown on (111)-oriented CZT wafer, non-oriented CZT wafer and FTO substrates is 3.4×10^{-2} lin^{-2}m^{-4}, 4.1×10^{-2} lin^{-2}m^{-4}, 4.9×10^{-2} lin^{-2}m^{-4}, respectively. The micro stress of CZT films grown on (111)-oriented CZT wafer is the lowest. The formula for dislocation density is as follows [25]:

$$\delta = \frac{1}{D^2} \tag{3}$$

According to the formula, the dislocation densities of CZT films grown on (111)-oriented CZT wafer, non-oriented CZT wafer and FTO substrates were calculated to be 1.91×10^{10} lines/m², 2.77×10^{10} lines/m², 3.81×10^{10} lines/m². When CZT films are grown directly on substrates with different lattice structures, the difference of lattice structures lead to the change of dislocation and stress in epitaxial films and affect the crystal quality and surface morphology of the films. For CZT films deposited on polished (111)-oriented CZT wafer, due to the homogeneous epitaxy and the lattice matching between the film and the substrate, the micro stress and the dislocation density of the film are decreased, and the crystal quality is improved. All the structural analysis shows that the properties of CZT films are affected by the substrate type during the deposition process.

3.2. Morphologies of CZT Films on Different Substrates

SEM images of CZT films grown on (111)-oriented CZT wafer, non-oriented CZT wafer and FTO substrates were also characterized. It can be seen from Figure 2 that the CZT films grown on (111)-oriented CZT wafer have more uniform particle sizes with triangle or quadrangle shapes, with the average particle sizes of about 10×20 μm². However, the particle sizes of CZT films grown on FTO and non-oriented CZT wafer have poor uniformity, which is due to the large dislocation densities between CZT films and the substrates (FTO and non-oriented CZT wafer). The surface composition ratios of Te: Zn: Cd were measured by XPS. The surface composition ratios of Te:Zn:Cd are 46.5:4.64:48.86, 45.5:3.84:50.67, 47.06:4.44:48.5 for CZT films grown on (111)-oriented CZT wafer, non -oriented CZT wafer and FTO substrates. The EDS mapping of CZT films on (111)-oriented CZT was also measured, as shown in Figure 3. The elements of Cd, Zn, Te are distributed uniformly at the surface of CZT films. The element of O is also detected. The atomic compositional ratios of Te:Zn:Cd:O are 43.7:4.9:50.5:0.9. The compositions of CZT films determined by EDS have similar results with that determined by XPS. The presence of oxygen has some possible reasons, such as introduced during the deposition process or deposited CZT films were oxidized in the atmosphere.

The morphology and crystal structure of CZT films were further analyzed by TEM and HRTEM. Figure 4a shows the TEM image of CZT. Figure 4b,c show HRTEM images of CZT films, respectively. Lattice images of CZT films are clearly presented and no crystal defects are observed. The interplanar spacing of 0.37 nm is consistent with surface (111) of CZT films. Figure 4d,e are the mapping images of CZT films, which show that Cd, Zn, Te elements coexist and are uniformly dispersed on CZT films.

Figure 2. *Cont.*

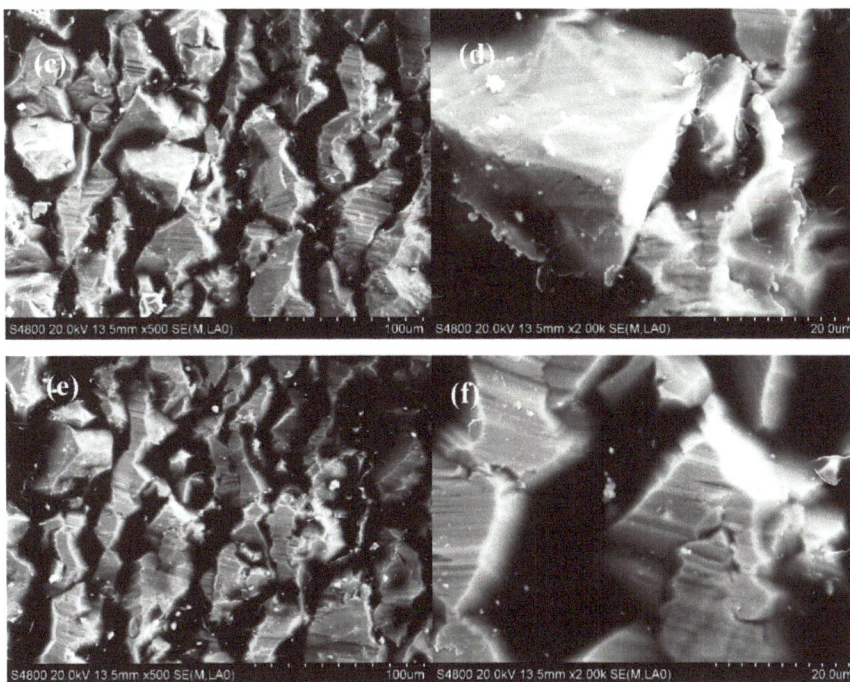

Figure 2. SEM characterizations of CZT films deposited on different substrates: (**a**,**b**) (111)-oriented CZT wafer, (**c**,**d**) non-oriented CZT wafer and (**e**,**f**) FTO substrates.

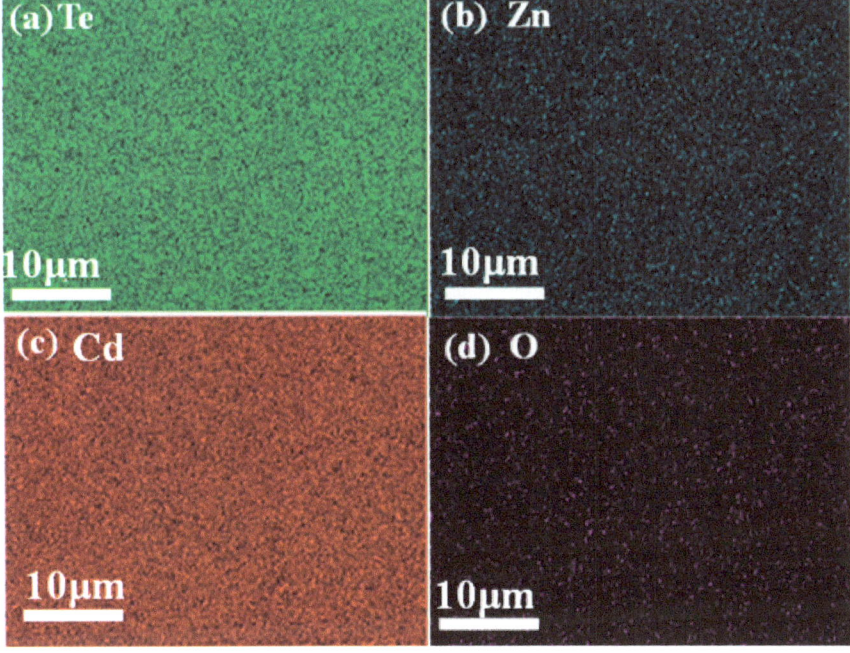

Figure 3. EDS mapping of CZT films on (111)-oriented CZT single crystal: (**a**) Te; (**b**) Zn; (**c**) Cd; (**d**) O.

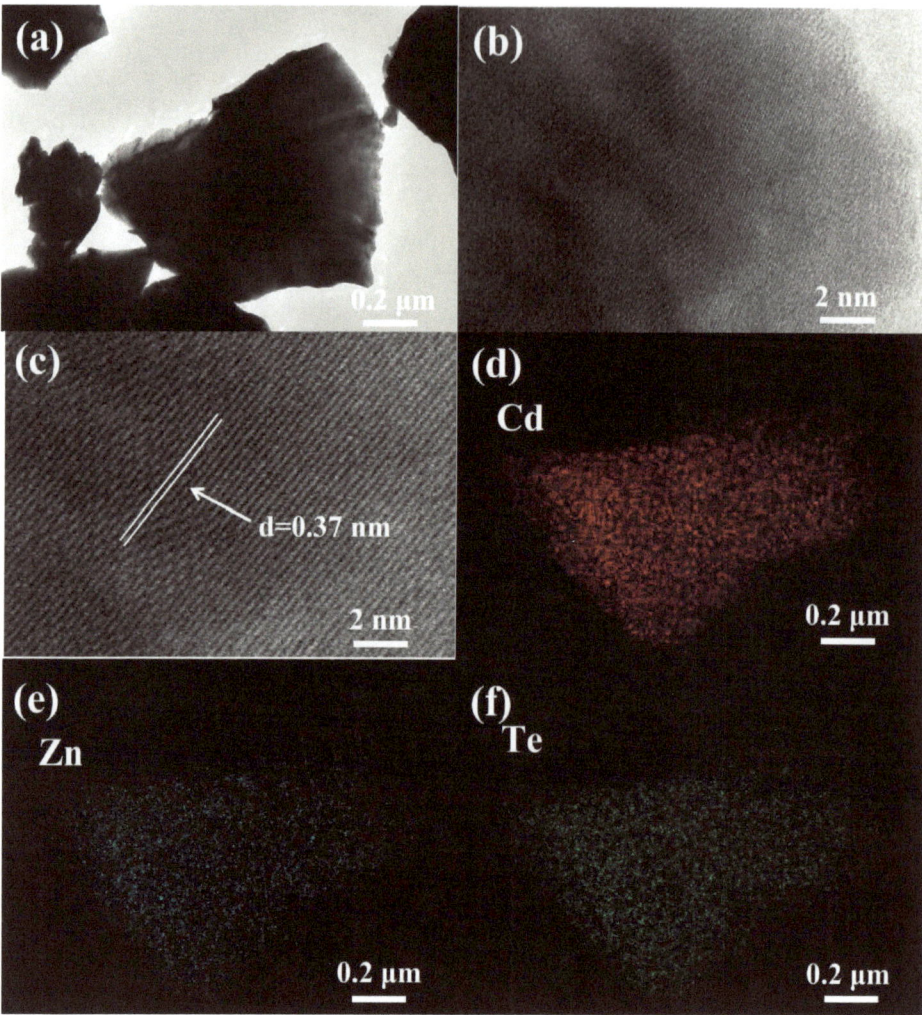

Figure 4. TEM (**a**) and HRTEM (**b,c**) images of CZT films; (**d–f**) Element mappings of Cd, Zn, Te.

3.3. Surface Treatment of CZT Films

CZT films have similar properties to those of CZT single crystal. Thus, in order to make the surface of CZT films smooth, the electroless $AuCl_3$ electrode can be deposited on the surface of CZT films. The surface treatment was carried out on CZT films to ensure the good surface properties of CZT film, which was first polished (PM6, Logitech) using Al_2O_3 with particle sizes of 0.3 μm and 0.05 μm for 10 min, and chemically polished using Br_2 + lactic acid + ethylene glycol for 2 min. Figure 5a shows the surface roughness of CZT films after mechanical polishing and Figure 5b shows the surface roughness of CZT films after chemical etching. It is found that R_a = 12.0 nm and R_q = 14.2 nm for mechanical polished CZT films, and the roughness reaches the normal value of CZT crystal after mechanical polishing. After chemical etching, R_a and R_q of CZT films in AFM are 21.4 nm and 26.8 nm, respectively. The reason for the increase of the roughness should be that the CZT films after etching exposes the air and the surface is oxidized, which has a certain impact on the roughness of the CZT films [26].

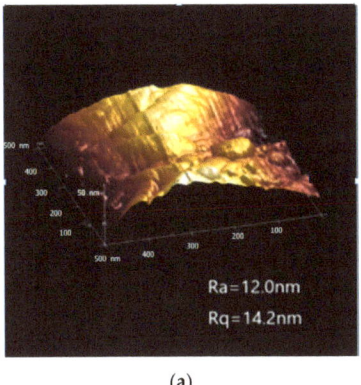 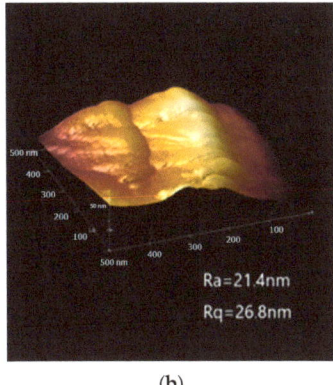

Figure 5. AFM characterizations of polished CZT films on (111)-oriented CZT wafer: (**a**) mechanical polishing, (**b**) chemical etching.

3.4. Electrical Properties of CZT Films

CZT films on different substrates were mechanically and chemically polished to ensure a smooth surface without a damage layer. Then, the electroless Au contact was deposited on the films, and the surface leakage currents of the films were measured. It can be seen from Figure 6 that the CZT films grown on (111)-oriented CZT wafer have a nearly linear I-V curve and the resistance of $\Delta V/\Delta I$ is about 9.4×10^9 Ω, which indicates that the films have a good quality and the lowest leakage current. CZT films grown on non-oriented CZT wafer have relatively linear I-V curves and the resistance of $\Delta V/\Delta I$ is about 9.1×10^9 Ω. However, CZT films grown on FTO substrate do not form good ohmic contact, with resistance of $\Delta V/\Delta I = 2 \times 10^9$ Ω. As demonstrated by the XRD and SEM measurements, CZT films deposited on difference substrates have different film qualities and surface morphologies, which will lead to different I-V characteristics of CZT films. CZT films on (111)-oriented CZT wafer have high crystal quality and a smooth and uniform surface, which results in better I-V properties. The resistance of CZT films has also some relation with the compositional ratios of CZT. Relatively high zinc content will enhance the resistance of CZT films, which is also contributive to improving the performance of CZT nuclear detectors. In our experiment, CZT films grown on non-oriented CZT wafer have a relatively high zinc content, which indicates that they are more suitable to fabricate CZT nuclear detectors. The carrier mobility and lifetime product ($\mu\tau$) of CZT films can be measured by a DC photoconductivity experiment [27]. Using a 650 nm laser as the light source, the I-V curves are measured by Keithley 2400. Then, the $\mu\tau$ product of carriers (electron or hole) can be obtained by the Hecht equation as follows: [28]

$$I(U) = I_0 \frac{\mu\tau U}{L^2}(1 - e^{-\frac{L^2}{\mu\tau U}}) \tag{4}$$

Here I_0 is the saturation photocurrent, U is the applied voltage, L is the film thickness. From Figure 7 that, it is revealed that the $\mu\tau$ value for the CZT films on (111)-oriented CZT wafer fitted by Hetch equation is 7.3×10^{-3} cm^2/V, which is in the range of $10^{-3} \sim 10^{-4}$ cm^2/V. Our detectors made from CZT films on (111)-oriented CZT wafer have similar $\mu\tau$ values compared to those of CZT single crystal, which indicates that CZT films on (111)-oriented CZT wafer have good carrier transport performance and have great potential for radiation detectors [29,30].

The energy spectrum of the high-resistance CZT films deposited on (111) oriented CZT single crystal was studied. By using [241]Am as the radiation source and putting the radiation source and CZT films detectors in the shielding box, its response to nuclear radiation was measured. The detectors were connected to an ORTEC 142IH preamplifier, and then to a Trump-PCI-2K multichannel analyzer system. Its pulse signal was observed on the oscilloscope and then the energy spectrum was measured. Figure 8 shows the nuclear radiation pulse signal of semi-insulating CZT films with the oscilloscope under ^{241}Am irradiation. After fitting, the pulse height is 2.7 V, FWHM is 1.27 µs, and baseline drift is −0.24 V. Figure 9 shows the energy spectrum of the CZT films detectors grown on (111)-oriented CZT wafer. Although the detection efficiency is not high, the energy spectrum can appear clearly. The FWHM (Full-width-at-half-maximum) of the energy spectrum is about 25%. It indicates that the CZT films on (111)-oriented CZT wafer have an acceptable response under irradiation of ^{241}Am. Due to the poor electrical properties of CZT films on non-oriented CZT wafer and FTO substrates, their photoconductivity cannot be measured exactly and correctly. The energy spectrum response and single pulse signal of nuclear radiation were also not detected.

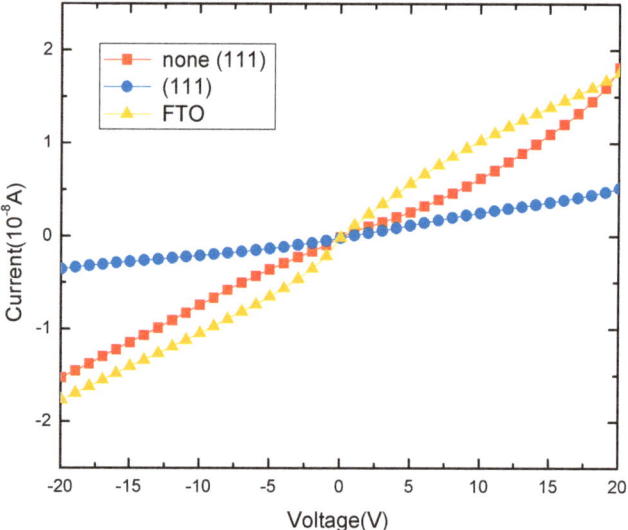

Figure 6. I-V curves of CZT films deposited on different substrates.

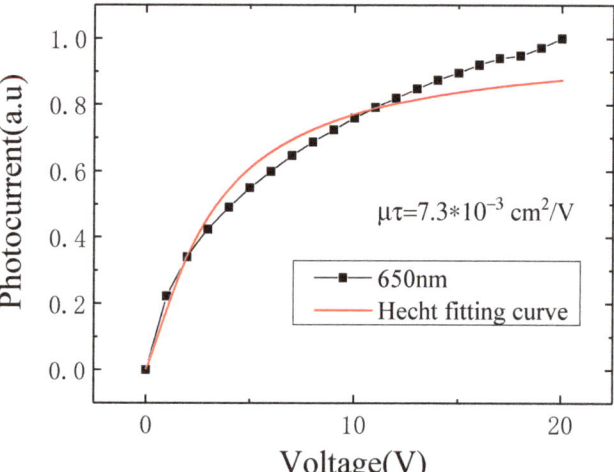

Figure 7. Photoconductivity curves of CZT films on (111)-oriented CZT wafer under 650 nm visible light.

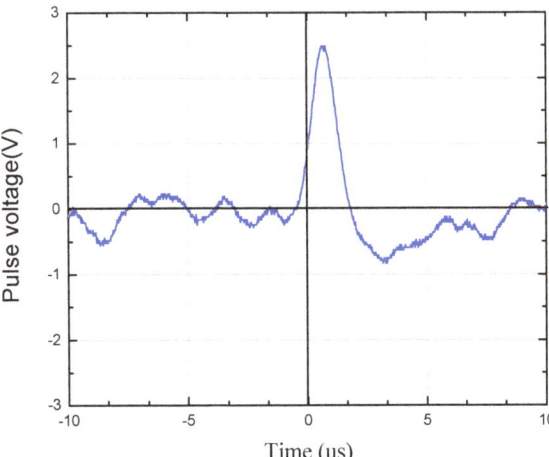

Figure 8. Single pulse signal of nuclear radiation on oscilloscope of CZT films grown on (111)-oriented CZT wafer.

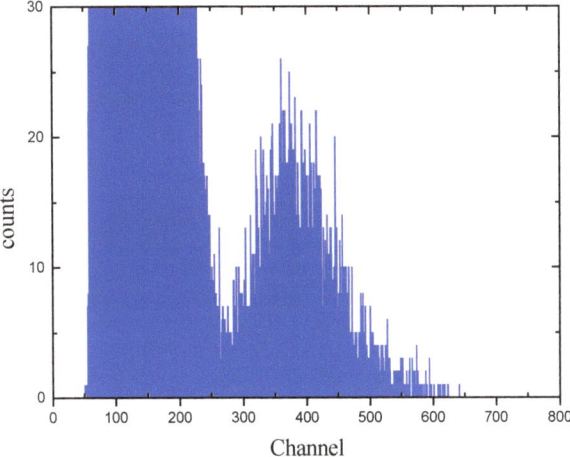

Figure 9. Energy spectrum response of CZT film on (111)-oriented CZT wafer under [241]Am irradiation.

4. Conclusions

In this paper, CZT films were grown on (111)-oriented CZT wafer, non-oriented CZT wafer and FTO substrates by CSS method. The effect of substrate type on the morphological, structural and electric properties of CZT films was studied. XRD measurements indicate that the diffraction peak of CZT films on (111)-oriented CZT wafer is the strongest, which indicates a high crystal quality. SEM and AFM measurements show that the surface of CZT films have been well modified, which contributes to forming good ohmic contact with electroless Au. CZT films grown on non-oriented CZT wafer have relatively high zinc content, which leads to their high resistance. I-V and DC photoconductivity measurements confirm that the electrical performance of CZT films on (111)-oriented CZT wafer have the resistance and $\mu\tau$ values of 9.4×10^9 Ω and 7.3×10^{-3} cm^2/V, respectively. CZT films grown on (111)-oriented CZT wafer have good reactions to nuclear radiation signals and can detect the radiation from weak radiation sources. Therefore, CZT films grown on (111)-oriented CZT wafer is suitable for fabricating nuclear detection devices.

Author Contributions: Investigation and Data curation: B.Y. and C.X.; Writing—original draft preparation: B.Y.; Validation: C.X.; Formal analysis and Visualization: M.X. and Y.J.; Administration: M.C. and J.Z.; Acquisition: M.C.; Conceptualization: J.Z.; Supervision: M.C. and L.W.; Project administration and Funding acquisition: L.W. All authors have read and agreed to the published version of the manuscript.

Funding: This research was funded by open topic of the State Key Laboratory of Nuclear Power Safety Monitoring Technology and Equipment [grant number K-A 2019.418]; open topic of Key Laboratory of Infrared Imaging Materials and Devices [grant number IIMDKFJJ-20-01].

Institutional Review Board Statement: Not applicable.

Informed Consent Statement: Not applicable.

Conflicts of Interest: The authors declare no conflict of interest.

References

1. Nemirovsky, Y.; Ruzin, A.; Asa, G.; Gorelik, Y.; Li, L. Study of Contacts to CdZnTe Radiation Detectors. *J. Electron. Mater.* **1997**, *26*, 756–764. [CrossRef]
2. Gao, X.Y.; Sun, H.; Yang, D.Y.; Wang, P.H.; Zhang, C.F.; Zhu, X.H. Large-area CdZnTe thick film based array X-ray detector. *Vacuum* **2021**, *183*, 109855. [CrossRef]
3. Pekarek, J.; Belas, E.; Zazvorka, J. Long-Term Stable Surface Treatments on CdTe and CdZnTe Radiation Detectors. *J. Electron. Mater.* **2017**, *46*, 1996–2002. [CrossRef]
4. Chen, R.Z.; Shen, Y.; Li, T.S.; Huang, J.; Gu, F.; Liang, X.Y.; Cao, M.; Wang, L.J.; Min, J.H. Interface optimization of free-standing CdZnTe films for solar-blind ultraviolet detection: Substrate dependence. *Vacuum* **2021**, *193*, 110484. [CrossRef]
5. Chander, S.; Dhaka, M.S. Thermal annealing induced physical properties of electron beam vacuum evaporated CZT films. *Thin Solid Films* **2017**, *625*, 131–137. [CrossRef]
6. Chander, S.; Purohit, A.; Patel, S.L.; Dhaka, M.S. Effect of substrates on structural, optical, electrical and morphological properties of evaporated polycrystalline CZT films. *Physica E* **2017**, *89*, 29–32. [CrossRef]
7. Chander, S.; Dhaka, M.S. Time evolution to $CdCl_2$ treatment on Cd-based solar cell devices fabricated by vapor evaporation. *Sol. Energy* **2017**, *150*, 577–583. [CrossRef]
8. Schlesinger, T.E.; Toney, J.E.; Yoon, H.; Lee, E.Y.; Brunett, B.A.; Franks, L.; James, R.B. Cadmium zinc telluride and its use as a nuclear radiation detector material. *Mater. Sci. Eng. R. Rep.* **2001**, *32*, 103–189. [CrossRef]
9. Tokuda, S.; Kishihara, H.; Adachi, S.; Sato, T. Preparation and characterization of polycrystalline CZT films for large-area, high-sensitivity X-ray detectors. *J. Mater. Sci. Mater. Electron.* **2004**, *15*, 1–8. [CrossRef]
10. Manez, N.; Sun, G.C.; Samic, H.; Berjat, B.; Kanoun, N.; Alquié, G.; Bourgoin, J.C. Material optimization for X-ray imaging detectors. *Nucl. Inst. Methods Phys. Res.* **2006**, *A567*, 281–284. [CrossRef]
11. Bansal, A.; Rajaram, P. Electrochemical growth of CZT films. *Mater. Lett.* **2005**, *59*, 3666–3671. [CrossRef]
12. Mycielski, A.; Szadkowski, A.; Lusakowska, E.; Kowalczyk, L.; Domagala, J.; Bak-Misiuk, J.; Wilamowski, Z. Parameters of substrates-single crystals of ZnTe and $Cd_{1-x}Zn_xTe$ (x < 0.25) obtained by physical vapor transport technique (PVT). *J. Cryst. Growth.* **1999**, *197*, 423–426.
13. Cohen, K.; Stolyarova, S.; Amir, N.; Chack, A.; Beserman, R.; Weil, R.; Nemirovsky, Y. MOCVD growth of ordered $Cd_{1-x}Zn_xTe$ epilayers. *J. Cryst. Growth.* **1999**, *198–199*, 1174–1178. [CrossRef]
14. Alamri, S.N. The growth of CdTe thin film by close space sublimation system. *Phys. Status Solidi (a)* **2003**, *200*, 352–360. [CrossRef]
15. Wu, S.H.; Zha, G.Q.; Cao, K.; Fu, J.H.; Li, Y.; Wang, Y.W.; Jie, W.Q.; Tan, T.T. The growth of CdZnTe epitaxial thick film by close spaced sublimation for radiation detector. *Vacuum* **2019**, *168*, 108852. [CrossRef]
16. Littler, C.L.; Gorman, B.P.; Weirauch, D.F.; Liao, P.K.; Schaake, H.F. Temperature, thickness, and interfacial composition effects on the absorption properties of (Hg, Cd)Te epilayers grown by liquid-phase epitaxy on CZT. *J. Electron. Mater.* **2005**, *34*, 768–772. [CrossRef]
17. Huang, J.; Chen, Z.R.; Bie, J.Y.; Shang, Y.; Yao, K.F.; Tang, K.; Cao, M.; Wang, L.J. Novel CdZnTe micro pillar films deposited by CSS method. *Mater. Lett.* **2019**, *263*, 127277. [CrossRef]
18. Cao, K.; Jie, W.Q.; Zha, G.Q.; Hu, R.Q.; Wu, S.H.; Wang, Y.W. Improvement of crystalline quality of CdZnTe epilayers on GaAs (001) substrates with a two-step growth by Close Spaced Sublimation. *Vacuum* **2019**, *164*, 319–324. [CrossRef]
19. Gao, J.N.; Jie, W.Q.; He, Y.H.; Sun, J.; Zhou, H.; Zha, G.Q.; Yuan, Y.Y.; Tong, J.L.; Yu, H.; Wang, T. Study of Te aggregation at the initial growth stage of CdZnTe films deposited by CSS. *Appl. Phys. A* **2012**, *108*, 447–450. [CrossRef]
20. Cao, K.; Zha, G.P.; Zhang, H.; Wang, A.Q.; Li, Y.; Wan, X. Preparation of $Cd_{0.8}Zn_{0.2}Te/Cd_{0.5}Zn_{0.5}Te/n^+$-GaAs thick film radiation detectors by close spaced sublimation. *Vacuum* **2021**, *192*, 110426. [CrossRef]
21. Yang, F.; Huang, J.; Zou, T.Y.; Tang, K.; Zhang, Z.L.; Ma, Y.C.; Gou, S.F.; Shen, Y.; Wang, L.J.; Lu, Y.C. The influence of surface processing on properties of CdZnTe films prepared by close-spaced sublimation. *Surf. Coat. Technol.* **2019**, *357*, 575–579. [CrossRef]
22. Lalitha, S.; Sathyamoorthy, R.; Senthilarasu, S.; Subbarayan, A.; Natarajan, K. Characterization of CdTe thin film-dependence of structural and optical properties on temperature and thickness. *Sol. Energy Mater. Sol. Cells* **2004**, *82*, 187–199. [CrossRef]

23. Bell, S.J.; Baker, M.A.; Duarte, D.D.; Schneider, A.; Seller, P.; Sellin, P.J.; Veale, M.C.; Wilson, M.D. Characterization of the metal-semiconductor interface of gold contacts on CdZnTe formed by electroless depositon. *J. Phys. D Appl. Phys.* **2015**, *48*, 275304. [CrossRef]
24. Chandera, S.; Dhaka, M.S. Enhanced structural, electrical and optical properties of evaporated CdZnTe thin films deposited on different substrates. *Mater. Lett.* **2017**, *186*, 45–48. [CrossRef]
25. Vidhya, S.N.; Balasundaram, O.N.; Chandramohan, M. The effect of annealing temperature on structural, morphological and optical properties of CdZnTe thin films. *Optik* **2015**, *126*, 5460–5463. [CrossRef]
26. Zheng, Q.; Dierre, F.; Crocco, J.; Carcelen, V.; Dieguez, E. Influence of surface preparation on CdZnTe nuclear radiation detectors. *Appl. Surf. Sci.* **2011**, *257*, 8742–8746. [CrossRef]
27. Cui, Y.; Groza, M.; Wright, G.W.; Roy, U.N.; Burger, A.; Li, L.; Lu, F.; Black, M.A.; James, R.B. Characterization of $Cd_{1-x}Zn_xTe$ crystals grown from a modified vertical bridgman technique. *J. Electron. Mater.* **2006**, *35*, 1267–1274. [CrossRef]
28. Ling, Y.P.; Min, J.H.; Liang, X.Y. Carrier transport performance of $Cd_{0.9}Zn_{0.1}Te$ detector by direct current photoconductive technology. *J. Appl. Phys.* **2017**, *121*, 034502. [CrossRef]
29. Yücel, H.; Birgül, O.; Uyar, E.; Çubukçu, S. A novel approach in voltage transient technique for the measurement of electron mobility and mobility-lifetime product in CdZnTe detectors. *Nucl. Eng. Technol.* **2019**, *51*, 731–737. [CrossRef]
30. Sturm, B.W.; He, Z.; Zurbuchen, H.; Koehn, P.L. Investigation of the asymmetric characteristics and temperature effects of CdZnTe detectors. *IEEE Trans. Nucl. Sci.* **2005**, *52*, 2068–2075. [CrossRef]

Article

Evaluation of Relationship between Grain Morphology and Growth Temperature of HgI$_2$ Poly-Films for Direct Conversion X-ray Imaging Detectors

Gang Xu [1,*], Ming Yao [1], Mingtao Zhang [1], Jinmeng Zhu [1], Yongxing Wei [1], Zhi Gu [2] and Lan Zhang [3]

1. School of Materials and Chemical Engineering, Xi'an Technological University, Xi'an 710032, China; yaom1118@163.com (M.Y.); ZMT3141010@163.com (M.Z.); nanojin@xatu.edu.cn (J.Z.); YongxingWEI@126.com (Y.W.)
2. State Key Laboratory of Solidification Processing, Northwestern Polytechnical University, Xi'an 710072, China; gu_zhi@163.com
3. Nuctech Company Limited, Beijing 100084, China; ZL_yulongfeng@163.com
* Correspondence: xugang@xatu.edu.cn

Citation: Xu, G.; Yao, M.; Zhang, M.; Zhu, J.; Wei, Y.; Gu, Z.; Zhang, L. Evaluation of Relationship between Grain Morphology and Growth Temperature of HgI$_2$ Poly-Films for Direct Conversion X-ray Imaging Detectors. *Crystals* **2022**, *12*, 32. https://doi.org/10.3390/cryst12010032

Academic Editors: Dah-Shyang Tsai, Andrian Kuchuk and Alberto Girlando

Received: 18 November 2021
Accepted: 23 December 2021
Published: 26 December 2021

Publisher's Note: MDPI stays neutral with regard to jurisdictional claims in published maps and institutional affiliations.

Copyright: © 2021 by the authors. Licensee MDPI, Basel, Switzerland. This article is an open access article distributed under the terms and conditions of the Creative Commons Attribution (CC BY) license (https://creativecommons.org/licenses/by/4.0/).

Abstract: The relationship between depositing temperature and crystallinity of grain for HgI$_2$ polycrystalline film with 170 cm^2 in area deposited by Physical Vapor Deposition (PVD) was investigated, considering the matches with readout matrix pixelation for female breast examination. The different depositing temperatures, 35, 40 and 45 °C, were carried out with the same source temperature, 100 °C, corresponding to 2–2.5 h of the growth period. The films deposited were investigated by XRD, SEM, and I–V. The results show that the grain size of the films grown increases with the depositing temperature from 35 to 45 °C. At 45 °C, the polycrystalline film has a preferred microcrystal orientation with 97.2% of [001]/[hkl] and grain size is about 180–220 μm. A 256 × 256 pixels X-ray image of a bolt, key, and wiring displacement was present distinctly with 50 keV with 6 mA current of X-ray generator. Our discussions on the relationship between depositing temperature and crystallinity of grain of film suggest that the higher growth temperature, the better crystallinity and excellent preferred microcrystal orientation of grain, however, with complementary bigger grain size. For matching readout matrix pixelation, the growth period of poly-films would be reduced appropriately for reasonable grain size and preventing the crack of films deposited.

Keywords: mercuric iodide; polycrystalline films; depositing temperature; x-ray imager

1. Introduction

Among a variety of detector materials, such as CdTe [1,2], PbI$_2$ [3–5], BiI$_3$ [6,7], and HgI$_2$ [8–12], HgI$_2$ turned out to be the superb one for direct and digital X-ray imaging for the last 15 years. Up to now, there is still a surge of interest in large area mercuric iodide (HgI$_2$) polycrystalline films in the application of digital radiographic detectors for medical diagnostic, scientific, and industrial applications on room-temperature X-rays [11,13–15], especially the application of female breast lesion. Compared with PbI$_2$, HgI$_2$ imagers demonstrate much sharper images due to excellent spatial resolution [9], which is easier to clearly detect the nidus of female breast cancer. As a layered structure material, the band gap of HgI$_2$ is about 2.13 eV, which matches the low-energy X-ray [13]. More importantly, resistivity with the order of 10^{10} Ohm·cm would be considered adequate for detector-level films [13], which was easy to be acquired due to the commercialization of the high-purity Hg and I$_2$ raw material, as well as purification of HgI$_2$ powder [16,17]. Hence, some methods have been used for the growth of HgI$_2$ poly-film, such as vapor phase deposition, laser ablation, screenprint [18,19], casting technique [20], electrode deposition, and growth from solution [10].

Because of the high vapor pressure of HgI$_2$ below its melting point, especially vacuum growth circumstance (higher than or equal to 10^{-3} Pa) rejecting the introduction of impurities, Physical Vapor Deposition (PVD) was mainly adopted for the growth of large-area

HgI$_2$ films [11–13]. It is worth indicating that the growth of HgI$_2$ poly-film by PVD had the lowest source temperature, the lowest depositing temperature and the shortest period, compared with those of PbI$_2$, BiI$_3$, PbBr$_2$, HgBr$_2$, and HgBrI, which was thought for a future scale-up to larger areas [14]. Hence, the films have been grown with area from tens to hundreds cm^2 [21,22], and thickness from tens to 1800 μm [10,23–25] by PVD for the application of imaging detectors.

In earlier studies, we had reported the growth of HgI$_2$ polycrystalline film (36 cm^2) by PVD, and the imaging effect on the Thin Film Transistor (TFT) with 256 × 256 pixels [12]. Although the imaging effect was thought to match with the basic requirements of detectors, the texture of films for detectors approximates to that of the HgI$_2$ powder according to XRD pattern. Fornaro [25] claims that HgI$_2$ film matches readout matrix pixelation and is suitable for the needs of digital radiography when polycrystalline, while it yields maximum radiation absorption and appropriate surface for electrode deposition when oriented. Hence, the promising work is to choose a compromising process to meet both requirements above, considering our previous processes [12].

In this paper, we investigated the correlation between depositing temperature and morphology of grains of polycrystalline HgI$_2$ with the area of 13 × 13 cm^2 in area and >500 μm in thickness by PVD. XRD, SEM, and I–V were used to characterize the properties of these as-grown films. Moreover, the detector prepared using the film on Thin Film Transistor (TFT) with 256 × 256 pixels was investigated.

2. Materials and Methods

The synthesis method for high-purity raw material of HgI$_2$ and growth furnace for films have been previously reported in detail in reference [12]. ITO and liquid crystal TFT with 170 cm^2 in area were used as the substrate to grow the films, which was rinsed only in 18 MΩ de-ionized H$_2$O. The whole growth cavity in the furnace was sealed below a vacuum of 6.6 × 10^{-3} Pa. The different depositing temperatures were 35, 40 and 45 °C, respectively. The source temperature was 100 °C. The growth period was set to be 2–2.5 h.

The texture of poly-films was studied by SHIMADZU 6000 X-ray diffraction meter with the X-ray wavelength of 1.5369 Å. The surface morphology of the films was examined using a FEI Quanta 400FScanning Electron Microscope. I–V characteristic of the as-grown film was measured by an Agilent 4155C IV instrument at room temperature. The digital X-ray imaging was obtained at Nuctech Company Limited, Beijing, China.

3. Results

The calculation method of growth temperature of some crystals in vapor has been provided [26]. Xu [27] optimized the growth temperature of HgI$_2$ single crystal and poly-crystal films, as shown in Equation (1), and thought that the depositing temperature would increase with the increase of source temperature in a seal system. At the same time, the reasonable depositing temperature is 78~130 °C (0.3~0.5 Tm, Tm: Melting point, 259 °C for HgI$_2$) for polycrystalline of HgI$_2$, which is favorable to the formation of crystallography facets for grain, with complementary rectangular columnar structure [27]. Considering the match with readout matrix pixelation, the choice of temperature should be close to the lower limit for tiny grains.

$$T_{min}^{3/2} exp(-\frac{E_{SD}}{KT_{min}}) = 0.085 \alpha_C a^2 M^{-1/2} \Delta P \tag{1}$$

Figure 1 gives the top and side of SEM photos of three different depositing temperature films with 170 cm^2 area. For Figure 1a,b, the corresponding depositing temperature is 35 °C with 2.5 h in growth period. The grain size is less than 50 μm, and the thickness is about 1200 μm. Many grains agglomerate so the grain size cannot be distinctly observed in Figure 1a. Meanwhile, the films look like a stack of small HgI$_2$ powder, viewing along the direction of growth from Figure 1b. Only on the top of the films, a few crystallographic planes are presented. Obviously, lower depositing temperature did not provide adequate energy for horizontal diffusion of sublimed HgI$_2$ molecular on ITO, restricted the form of the crystallographic planes. Hence, the orientation of this film is not a very reasonable result, although detector assembled with this kind of film is available [12]. The grain size is about 120–180 µm with the thickness of 480–500 µm, as shown in Figure 1c,d, corresponding to 40 °C of depositing temperature and 2 h of growth period. Although the bottom of the films is also close to the stacking of the powders, the top part presents better crystalline morphology, which is an evident issue of raising the depositing temperature. The best effect was shown in Figure 1e,f for 45 °C of deposition, accompanied with 2 h in growth period. The films with grain size of 180–220 µm and thickness of 520 µm demonstrate a preferred microcrystal orientation, with complementary rectangular facets of the grains, and appropriate surface for electrode deposition. It is apparent that the result in Figure 1e,f is more adaptive for the necessity of assembling detector.

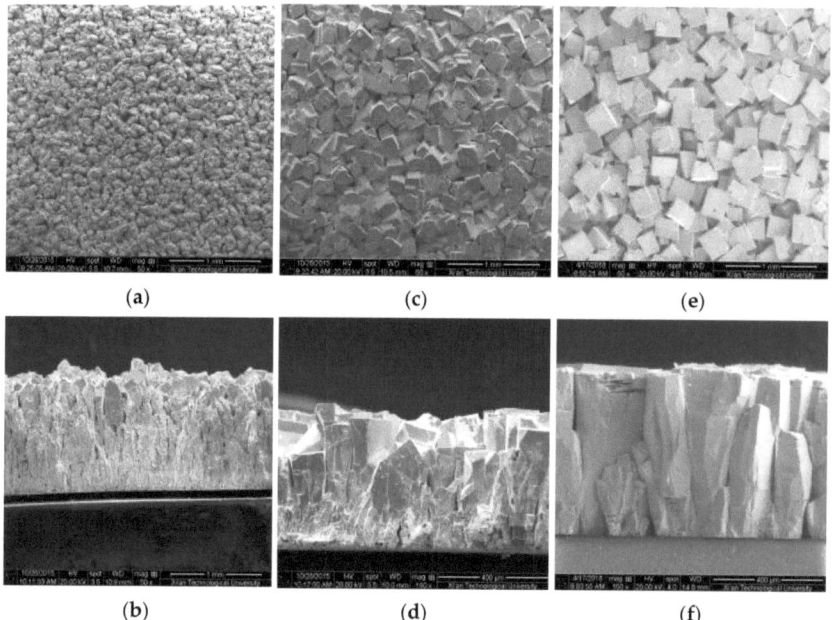

Figure 1. Top (**a,c,e**) and side (**b,d,f**) SEM views of polycrystalline HgI$_2$ films.

Figure 2 shows a HgI$_2$-coated 13 × 13 cm^2 on ITO with 45 °C of the depositing temperature and 100 °C of the source temperature, as well as 2 h growth period. It is obvious that red film distributes homogeneously. Figure 3 gives the XRD characteristic of the above film. The main (001) peak is very clear. According to the expressions $\Sigma I(00l)/\Sigma I$ (hkl), the orientation along [001] was evaluated to be approximately 97.2%, which indicated a very excellent texture, while the texture of the film in Figure 1a,b could be estimated to close that of HgI$_2$ powder according to the previous work [12].

Figure 2. HgI$_2$ -coated 13 × 13 cm^2 on ITO.

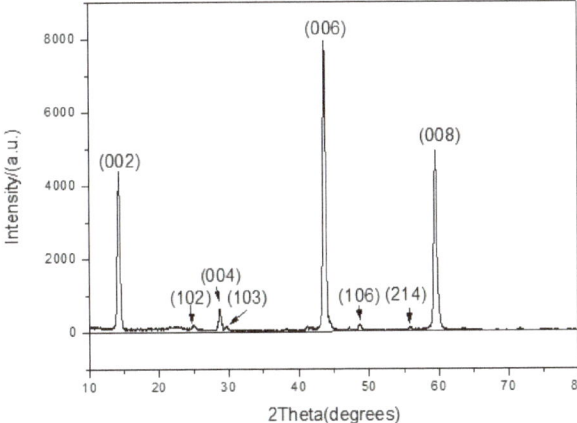

Figure 3. XRD pattern of polyscrystalline HgI$_2$ film.

The film's growth on TFT corresponding to the best growth process, grown at 45 °C for 2 h was carried out. Au was sputtered onto the HgI$_2$ poly-films as top electrode and TFT as back one, forming Au/HgI$_2$/TFT contacts then. I–V curves of HgI$_2$ films were measured using Agilent 4155C IV instrument at 25 °C. The testing result is shown in Figure 4. We can see the I–V curves present an approximate linear relation. The resistivity of the films was calculated to be 0.9×10^{12} Ω·cm, which is coincident with the result published in reference [12] and also adequate for the need of detector-level films.

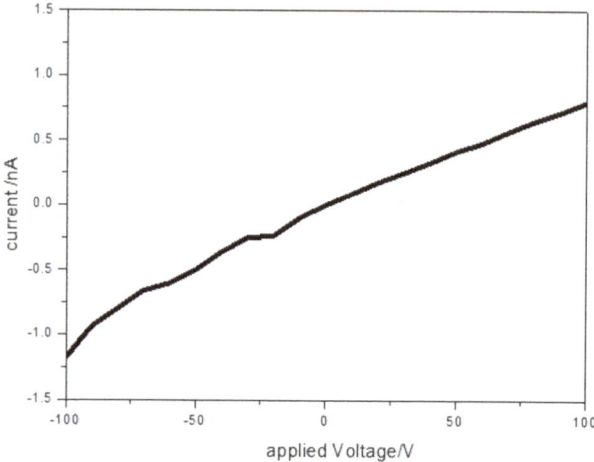

Figure 4. I–V characteristic of polyscrystalline HgI_2 film.

The imaging testing was carried out by the researcher in Nuctech Company Limited, Beijing. X-ray generator was set to 50 keV with 6 mA current. A key, a bolt with Φ 6 mm, and a wiring displacement were the objectives for imaging. Figure 5 shows planar X-ray images from 256 × 256 pixel HgI_2 poly-films for the above subjects. The profile of the key was very clear, and about 1 mm of thread spacing of the bolt can be distinctly presented. Especially, 16 threads with 0.3 mm in diameter of a wiring displacement can be accounted distinctly. Obviously, the result was superior to the previous testing [12].

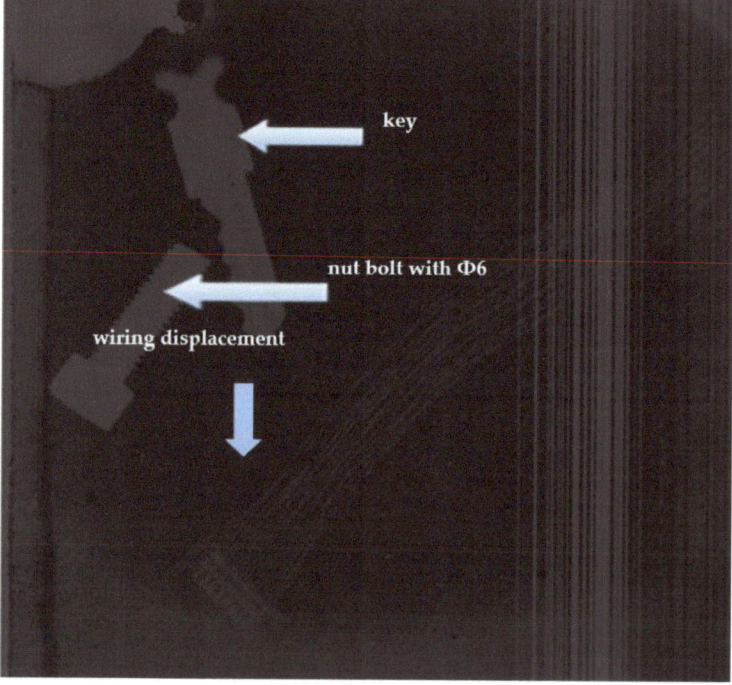

Figure 5. Bolt, key, and wiring displacement image taken by X-ray Planar Images.

Considering the present and previous work [12], we thought that three different kinds of texture of the films were all adequate for detector assembling, and the third deposition process should be the best one. As shown in Figure 1f, the surface roughness of the film gets better with the growth temperature arising, which is in favor of electrode deposition, comparing the former two surface structures of the films grown. More importantly, the improvement of the orientation due to the increasing of depositing temperature is favorable to the sensitivity of the detector [25]. Therefore, the present results attribute to (001) orientation of the film and formation of the uniform electrical field. Nevertheless, the crack presents on the top of the grains in Figure 1f. It is well known that HgI_2 is built from three-layer packages (I-Hg-I) with weak van der Walls bonding between adjacent planes of iodine atoms and perpendicular to the c-axis. This layered structure determines a high anisotropy for most of the properties. Isshiki [28] reported that HgI_2 has indeed rather low thermal conductivity, 4 and 20 mW/cm·K, for the crystal direction parallel and perpendicular to the c-axis of HgI_2 crystal. This is the reason why the crack occurs likely along the direction of parallel, other than perpendicular to (001) of crystal. The temperature of the growth surface arises due to the low thermal conductivity of HgI_2 during crystal growth [29], leading to thermal stress on the top of the columnar grain and crack occurs. Meanwhile, the decline of supersaturation in vapor due to the increasing depositing temperature leads to the growth driving force decrease with time. Thereby, the thickness of the film attenuated to some extend with the elevation of depositing temperature in this work. Furthermore, the actual temperature on the surface of film would rise above one measured in the depositing field of ploy-films considering the thermal conductivity of crystal, which would be responsible for larger grain size shown in Figure 1c,e, especially the occurrence of cracks on the top of grain in Figure 1f. The above issues indicate the growth period would be reduced appropriately. Therefore, further study would focus on the growth of grain size less than 100 μm and the better orientation.

4. Conclusions

A large area HgI_2 poly-films (170 cm^2) was grown by PVD, applying for female breast examination. A 256 × 256 pixels X-ray image of a bolt, key, and a wiring displacement was distinctly presented with 50 keV with 6 mA current of X-ray generator. For HgI_2 poly-films (170 cm^2), films depositing at 35 °C look like the stack of HgI_2 powder, present better crystalline morphology at 40 °C, and an excellent microcrystal orientation with 97.2% of [001]/[hkl] at 45 °C. The relationship between depositing temperature and crystallinity of grain shows that the higher depositing temperature, the better crystallinity and excellent preferred microcrystal orientation of grain, the larger the grain size. For matching readout matrix pixelation, the growth period of HgI_2 poly-films poly-films would be reduced for reasonable grain size and preventing the formation of grain cracks.

Author Contributions: Conceptualization, L.Z. and G.X.; methodology, G.X. and Z.G.; investigation, L.Z. and G.X.; resources, L.Z. and Z.G.; data curation, M.Y., M.Z. and G.X.; validation, J.Z. and Y.W.; writing—original draft preparation, G.X. and Z.G.; project administration, G.X. and L.Z. All authors have read and agreed to the published version of the manuscript.

Funding: This research was funded by the National Natural Science Foundation of China, grant No. 52002304, National Natural Science Foundation of China, grant No. 11704301.

Institutional Review Board Statement: Not applicable.

Informed Consent Statement: Not applicable.

Data Availability Statement: Available on request.

Conflicts of Interest: The authors declare no conflict of interest.

References

1. Cha, B.K.; Yang, K.; Cha, E.S.; Yong, S.M.; Heo, D.; Kim, R.K.; Jeon, S.; Seo, C.W.; Kim, C.R.; Ahn, B.T.; et al. Structural and electrical properties of polycrystalline CdTe films for direct X-ray imaging detectors. *Nucl. Instr. Meth. A* **2013**, *731*, 320–324. [CrossRef]
2. Brus, V.V.; Solovan, M.N.; Maistruk, E.V.; Kozyarskii, I.P.; Maryanchuk, P.D.; Ulyanytsky, K.S.; Rappich, J. Specific features of the optical and electrical properties of polycrystalline CdTe films grown by the thermal evaporation method. *Phy. Solid State* **2014**, *56*, 1947–1951. [CrossRef]
3. Zhu, X.-H.; Sun, H.; Yang, D.-Y.; Yang, J.; Li, X.; Gao, X.-Y. Fabrication and characterization of X-ray array detectors based on polycrystalline PbI_2 th ick films. *J. Mater. Sci. Mater. Electron.* **2014**, *25*, 3337–3343.
4. Zhu, X.H.; Sun, H.; Yang, D.Y.; Zheng, X.L. Growth, surface treatment and characterization of polycrystalline lead iodide thick films prepared using close space deposition technique. *Nucl. Instr. Meth. A* **2012**, *691*, 10–15. [CrossRef]
5. Mulato, M.; Condeles, J.F.; Ugucioni, J.C. Comparative Study of HgI_2, PbI_2 and TlBr Films Aimed for Ionizing Radiation Detection in Medical Imaging. *Mrs Online Proc. Libr.* **2011**, *1341*, 145–150. [CrossRef]
6. Fornaro, L.; Saucedo, E.; Mussio, L.; Gancharov, A.; Cuña, A. Bismuth tri-iodide polycrystalline films for digital X-ray radiography applications. *IEEE Trans. Nucl. Sci.* **2004**, *51*, 96–100. [CrossRef]
7. Cuña, A.; Aguiar, I.; Gancharov, A.; Pérez, M.; Fornaro, L. Correlation between growth orientation and growth temperature for bismuth tri-iodide films. *Cryst. Res. Technol.* **2004**, *39*, 899–905. [CrossRef]
8. Jiang, H.; Zhao, Q.; Antonuk, L.E.; Elmohri, Y.; Gupta, T. Development of active matrix flat panel imagers incorporating thin layers of polycrystalline HgI_2 for mammographic x-ray imaging. *Phys. Med. Biol.* **2013**, *58*, 703–714. [CrossRef]
9. Zentai, G.; Partain, L.; Pavlyuchkova, R.; Proano, C.; Virshup, G.; Melekhov, L.; Zuck, A.; Breen, B.N.; Dagan, O.; Vilensky, A.; et al Mercuric iodide and lead iodide x-ray detectors for radiographic and fluoroscopic medical imaging. *Proc. SPIE* **2003**, *5030*, 77–91
10. Iwanczyk, J.S.; Patt, B.E.; Tull, C.R.; MacDonald, L.R.; Skinner, N.; Fornaro, L. HgI_2 polycrystalline films for digital X-ray imagers. *IEEE Trans. Nucl. Sci.* **2002**, *49*, 160–164. [CrossRef]
11. Iwanczyk, J.S.; Patt, B.E.; Tull, C.R.; MacDonald, L.R.; Skinner, N.; Hoffman, E.J.; Fornaro, L.; Mussio, L.; Saucedo, E.; Gancharov, A. Mercuric iodide polycrystalline films. *Proc. SPIE* **2001**, *4508*, 28–41.
12. Xu, G.; Li, J.Y.; Nan, R.H.; Zhou, W.L.; Gu, Z.; Zhang, L.; Ma, X.M.; Cao, X.P. Large area HgI_2 polycrystalline films for X-Ray imager. *J. Optoelectron. Adv. Mater.* **2016**, *18*, 842–846.
13. Burger, A.; Nason, D.; Franks, L. Mercuric iodide in prospective. *J. Cryst. Growth* **2013**, *379*, 3–6. [CrossRef]
14. Katsis, A.; Abel, E.; Pavlyuchkova, R.; Senapati, S.; Girdhani, S.; Parry, R. Abstract 3599: Implementation of a fully automated 3D foci counting algorithm to determine DNA damage in cells. *Cancer Res.* **2016**, *76*, 3599.
15. El-Mohri, Y.; Antonuk, L.E.; Zhao, Q.; Su, Z.; Yamamoto, J.; Du, H.; Sawant, A.; Li, Y.; Wang, Y. TH-C-I-611-09: Development of Direct Detection Active Matrix Flat-Panel Imagers Employing Mercuric Iodide for Diagnostic Imaging. *Med. Phys.* **2005**, *32*, 2158. [CrossRef]
16. Piechotka, M. Mercuric iodide for room temperature radiation detectors. Synthesis, purification, crystal growth and defect formation. *Mater. Sci. Eng.* **1997**, *18*, 1–98. [CrossRef]
17. Van Scyoc, J.M.; James, R.B.; Burger, A.; Soria, E.; Perrino, C.; Cross, E.; Schieber, M.; Natarajan, M. Electrodrift purification of mercuric iodide for improved gamma-ray detector performance. *Nucl. Instrum. Methods Phys. Res. Sect. A Accel. Spectrometers Detect. Assoc. Equip.* **1996**, *380*, 36–41. [CrossRef]
18. Schieber, M.; Zuck, A.; Braiman, M.; Melekhov, L.; Nissenbaum, J.; Turchetta, R.A.D.; Dulinski, W.; Husson, D.; Riester, J.L Polycrystalline mercuric iodide detectors. *Proc. SPIE* **1997**, *3115*, 146–151.
19. Schieber, M.; Zuck, A.; Melekhov, L.; Shatunovksy, R.; Hermon, H.; Turchetta, R.A.D. High-flux x-ray response of composite mercuric iodide detectors. *Proc. SPIE* **1999**, *3768*, 296–310.
20. Ugucioni, J.C.; Netto, T.G.; Mulato, M. Mercuric iodide composite films using polyamide, polycarbonate and polystyrene fabricated by casting. *Nucl. Instr. Meth. A* **2010**, *622*, 157–163. [CrossRef]
21. Xu, G.; Guo, Y.F.; Xi, Z.Z.; Gu, Z.; Zhang, L.; Yu, W.T.; Ma, X.M.; Li, B. Study on growth of large area mercuric iodide polycrystalline film and its X-ray Imaging. *Proc. SPIE* **2014**, *9298*, 91–97.
22. Hartsough, N.E.; Iwanczyk, J.S.; Nygard, E.; Malakhov, N.; Barber, W.C.; Gandhi, T. Polycrystalline Mercuric Iodide Films on CMOS Readout Arrays. *IEEE Trans. Nucl. Sci.* **2009**, *56*, 1810–1816. [CrossRef]
23. Schieber, M.; Hermon, H.; Zuck, A.; Vilensky, A.; Melekhov, L.; Shatunovsky, R.; Meerson, E.; Saado, Y.; Lukach, M.; Pinkhasyc, E.; et al. Thick films of X-ray polycrystalline mercuric iodide detectors. *J. Cryst. Growth* **2001**, *225*, 118–123. [CrossRef]
24. Roy, U.N.; Cui, Y.; Wright, G.; Barnett, C.; Burger, A.; Franks, L.A.; Bell, Z.W. Polycrystalline mercuric iodide films: Deposition, properties, and detector performance. *IEEE Trans. Nucl. Sci.* **2002**, *49*, 1965–1967. [CrossRef]
25. Fornaro, L. State of the art of the heavy metal iodides as photoconductors for digital imaging. *J. Cryst. Growth* **2013**, *371*, 155–162 [CrossRef]
26. Dryburgh, P.M. The estimation of minimum growth temperature for crystals grown from the gas phas. *J. Cryst. Growth* **1988**, *87*, 397–407. [CrossRef]
27. Xu, G.; Gu, Z.; Jian, Z.Y.; Xi, Z.Z.; Liu, C.X.; Zhang, G. Study on the lowest growth temperature of mercuric iodide. *J. Synt. Growth* **2012**, *41*, 1195–1199.

28. Isshiki, M.; Piechotka, M.; Kaldis, E. Vapor growth kinetics of α-HgI_2 crystals. *J. Cryst.Growth* **1990**, *102*, 344–348. [CrossRef]
29. Zaletin, V.M.; Lyakh, N.V.; Ragozina, N.V. Stability of growth conditions and α-HgI_2 crystal habit during growing by temperature oscillation method. *Cryst. Res. Technol.* **1985**, *20*, 307–312. [CrossRef]

Article

Effect of Irradiation with Low-Energy He²⁺ Ions on Degradation of Structural, Strength and Heat-Conducting Properties of BeO Ceramics

Maxim V. Zdorovets [1,2,3], Dmitriy I. Shlimas [1,2], Artem L. Kozlovskiy [1,4,*] and Daryn B. Borgekov [1,2]

[1] Laboratory of Solid State Physics, The Institute of Nuclear Physics, Almaty 050032, Kazakhstan; mzdorovets@inp.kz (M.V.Z.); shlimas@mail.ru (D.I.S.); borgekov@mail.ru (D.B.B.)
[2] Engineering Profile Laboratory, L.N. Gumilyov Eurasian National University, Nur-Sultan 010008, Kazakhstan
[3] Department of Intelligent Information Technologies, Ural Federal University, 620075 Yekaterinburg, Russia
[4] Institute of Geology and Oil and Gas Business, Satbayev University, Almaty 050032, Kazakhstan
* Correspondence: kozlovskiy.a@inp.kz; Tel./Fax: +7-7024413368

Abstract: The paper is devoted to the study of radiation-induced damage kinetics in beryllium oxide ceramics under irradiation with low-energy helium ions with fluences of 10^{15}–10^{18} ion/cm². It was revealed that at irradiation fluences above 10^{17} ion/cm², a decrease in radiation-induced damage formation and accumulation rate is observed, which indicates the saturation effect. At the same time, the main mechanisms of structural changes caused by irradiation at these fluences are amorphization processes and dislocation density increase, while at fluences of 10^{15}–10^{16} ion/cm², the main mechanisms of structural changes are due to the reorientation of crystallites and a change in texture, with a small contribution of crystal lattice distorting factors. It was discovered that the radiation-induced damage accumulation as well as an implanted helium concentration increase leads to the surface layer destruction, which is expressed in the ceramic surface hardness and wear resistance deterioration. It was determined that with irradiation fluences of 10^{15}–10^{16} ion/cm², the decrease in thermal conductivity is minimal and is within the measurement error, while an increase in the irradiation fluence above 10^{17} ion/cm² leads to an increase in heat losses by more than 10%.

Keywords: beryllium oxide; helium swelling; radiation defects; low-energy ions; inert matrices

Citation: Zdorovets, M.V.; Shlimas, D.I.; Kozlovskiy, A.L.; Borgekov, D.B. Effect of Irradiation with Low-Energy He²⁺ Ions on Degradation of Structural, Strength and Heat-Conducting Properties of BeO Ceramics. *Crystals* **2022**, *12*, 69. https://doi.org/10.3390/cryst12010069

Academic Editors: Linghang Wang and Gang Xu

Received: 19 December 2021
Accepted: 3 January 2022
Published: 5 January 2022

Publisher's Note: MDPI stays neutral with regard to jurisdictional claims in published maps and institutional affiliations.

Copyright: © 2022 by the authors. Licensee MDPI, Basel, Switzerland. This article is an open access article distributed under the terms and conditions of the Creative Commons Attribution (CC BY) license (https://creativecommons.org/licenses/by/4.0/).

1. Introduction

In the light of recent developments in nuclear power and alternative energy in energy-intensive countries, particular attention is being paid to the development of technologies using new types and concepts of nuclear fuel use. One such concept is the concept of replacing traditional fuel elements with inert matrices that use plutonium or americium instead of uranium [1–3]. The increased interest in these types of materials is due to the fact that, during their operation, actinides or plutonium are not formed, and the concentration of actinides in spent fuel is significantly lower than in conventional fuel elements or mixed oxide fuel [4–7]. At the same time, the transition of nuclear reactors to the use of plutonium fuel with an inert matrix that does not contain uranium makes it possible to increase the burnup of plutonium by a factor of 2–3, which is one of the main advantages of these materials. Generally, oxide ceramics such as ZrO_2, $MgAl_2O_4$, CeO_2, and BeO are used as inert matrix materials, interest in which is due to their high resistance to external influences, including heating and mechanical pressure or friction [8–10]. At the same time, one of the key problems in the use of inert nuclear fuel matrices in comparison with traditional fuel elements or mixed oxide fuel is the confirmation of the stability of these materials for long-term operation (more than 10–15 years) and the preservation of strength, mechanical, and heat-conducting properties [9,10].

One of the promising materials for inert oxide-based matrices is beryllium oxide (BeO) ceramics, which has a number of unique physicochemical, heat-conducting, and mechanical properties [11,12]. The interest in this class of materials is due to its inertness to most types of aggressive media and the ability to work at high temperatures, as well as good absorbing capacity and high thermal conductivity. All this allows us to use these ceramics as neutron reflectors or absorbers, various types of structural materials, etc. [13–15]. It is worth noting that despite the increase in interest in BeO ceramics in recent years, this type of ceramics has been used for a long time as a basis for dosimeters and OSL sensors designed to register ionizing radiation and control the dose load [16–19]. Interest in this area is due to high sensitivity and luminescence properties of beryllium oxide, which allows high accuracy of recording and further determination of radiation dose [16–19]. Moreover, an important area of research in radiation materials science is the study of the applicability and efficiency of doping with beryllium oxide or other rare earth elements of radiation-resistant ceramics and glasses [20–24]. These studies are based on an assessment of the possibility of an increase in the resistance of materials to radiation damage and accumulated defects in the structure, which can lead to disordering and deformation of the material. The main feature of doping in this case is the possibility of increasing stability due to substitution and interstitial processes, as well as the absorbing capabilities of rare earth elements or beryllium oxide, which leads to a slowdown in the accumulation of defects in the structure and subsequent deformation.

At the same time, as it was established earlier, ceramics based on beryllium oxide are highly resistant to radiation damage and the accumulation of radiation-induced defects, which opens up the possibility of their operation for a long time [25,26], while high absorption capacity indices make it possible to use them in fields of increased background radiation or large neutron fluxes.

One of the mechanisms of radiation-induced damage arising in materials used as inert matrices of nuclear fuel or structural materials is the accumulation of implanted or transmutation helium in the structure of the surface layer of ceramics [27–29]. The presence of poorly soluble and, at the same time, highly mobile helium in the structure of the surface layer can lead to its agglomeration, followed by the formation of gas-filled bubbles [30–32]. The formation of such bubbles in the structure of the surface layer of a ceramic or metal can lead to destructive processes of swelling and peeling of the surface layer, which in turn leads to destruction and deterioration of the properties of the material [33,34]. Furthermore, an important factor affecting the working and heat-conducting properties of ceramics is a decrease in the amount of heat removal from the system due to the destruction and destabilization of the properties of the heat-conducting material. The formation of distortions in the structure as a result of the accumulation of gas-filled bubbles can lead to a decrease in the thermal conductivity coefficient, which will lead to destabilization of heat removal from the system, as well as its overheating. From a mechanical point of view, the formation of additional distortions in the near-surface layer of ceramics can lead to a change in its strength and wear resistance, which also has a negative effect on the mode and time of operation.

At the same time, in the case of using oxide ceramics as materials of inert matrices or structural materials in nuclear power, the key factor in their application is knowledge of the kinetics and mechanisms of the degradation of structural, mechanical, and heat-conducting properties, depending on the degree of the accumulation of radiation-induced damage and the subsequent amorphization processes or disordering [35,36]. In this regard, the acquisition of any new data on radiation damage and their effect on change in the properties of ceramics is very significant and relevant today, which prompts a large number of scientific groups to engage in such studies. Based on the foregoing, the key goal of this article is to assess the effect of helium irradiation at doses of 10^{15}–10^{18} ion/cm^2 on the change in the mechanical, structural, and heat-conducting properties of BeO ceramics. Interest in this topic is not only due to the possibility of obtaining new data on the radiation resistance of these ceramics, but also due to the assessment of radiation-induced defects and the implanted helium concentration effect on thermal conductivity of ceramics. As is known from the literature, at doses above 10^{17} ion/cm^2 in the structure of oxide and nitride ceramics, the formation of gas-filled bubbles arising as a result of agglomeration and filling of helium-implanted voids in the ceramics structure is observed, which leads to its swelling and destruction [37–39]. However, there are still questions associated with the subsequent accumulation of helium when these radiation doses are exceeded, as well as a change in the rate of accumulation of defects and destruction of the material when the effect of saturation with defects occurs.

2. Experimental Part

Ceramics based on beryllium oxide obtained by hot pressing and having a high density (3.018 g/cm^3) close to the reference value (3.020 g/cm^3) were chosen as the samples under study. The original samples were purchased from a commercial company Berlox® (American Beryllia Inc., Haskell, NJ, USA) that is engaged in the production of ceramics for commercial and research purposes.

The samples were irradiated at a DC-60 heavy ion accelerator (INP ME RK, Nur-Sultan, Kazakhstan). Low-energy He^{2+} ions with an energy of 40 keV (20 keV/charge) at an ion flux density of 10^{10} ions/cm^2*s were used as incident ions. In order to avoid overheating of the samples during irradiation, special water-cooled target holders were used, which made it possible to maintain the temperature of the samples in the range of 30–50 °C, thereby excluding the effect of high-temperature annealing of defects during irradiation. The irradiation fluences were 10^{15}–10^{18} ion/cm^2, the choice of which is due to the possibility of simulating the structure swelling effects as a result of ion implantation and helium accumulation in the structure of the surface layer [37–39].

Figure 1 shows the simulation results of radiation damage and the implantation of helium ions in the structure of an irradiated ceramic layer with a depth of 400 nm. The maximum displacement value for fluences of 10^{17} ion/cm^2–10^{18} ion/cm^2 is 3–17 dpa, which, in comparison with radiation damage caused by neutron irradiation in the case of oxide ceramics, is 0.3–1.7 × 10^{22} neutron/cm^2. This atomic displacements value for the maximum irradiation fluences is due to the fact that the main contribution to the formation of radiation-induced defects is made by the energy losses of incident He^{2+} ions during interaction with nuclei, while the energy losses on the electron shells value is an order less than the nuclear losses value. According to calculations, the energy losses of incident ions on the nuclei are 182.4 keV/μm, and losses on electron shells are 10.7 keV/μm. In this case, the maximum damage accumulation area is at a depth of 170–300 nm from the surface of the samples, with a maximum at 250–170 nm. The maximum concentration of implanted ions is from 0.1 to 1.3 at.%, for fluences 10^{17} ion/cm^2–10^{18} ion/cm^2.

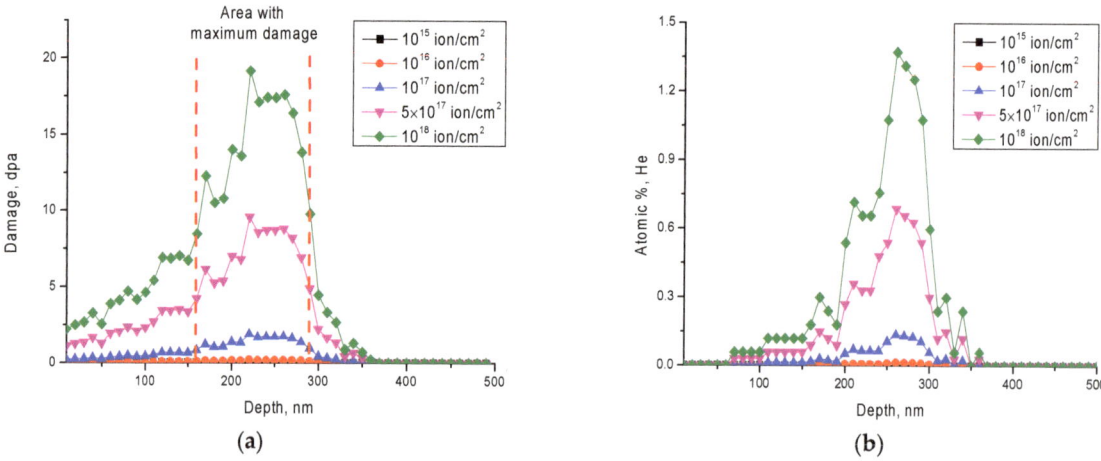

Figure 1. Simulation results for SRIM Pro 2013 [40,41]: (**a**) Damage vs. depth; (**b**) Atomic% He vs. depth.

The study of the effect of irradiation and accumulation of radiation-induced defects caused by helium ions on the structural properties and swelling of the crystal lattice was carried out by analyzing the X-ray diffraction patterns of the samples under study before and after irradiation. Diffraction patterns were obtained using a D8 ADVANCE ECO (Bruker, Berlin, Germany) powder diffractometer. Diffraction patterns were recorded in the Bragg-Brentano geometry in the angular range of $2\theta = 35$–$75°$, with a step of $0.03°$.

The hardness value was determined by the indentation method by using a standard method and using a Vickers pyramid at a load of 500 N. To determine the hardness value and its change as a result of irradiation and radiation-induced defects accumulation, 25 points were measured, which made it possible to determine the standard deviation of the hardness parameters.

The softening degree (SD) was determined from the change in the hardness of the near-surface layer before (H_0) and after (H) irradiation, determined by the indentation method using the calculation Equation (1):

$$SD = \left(\frac{H_0 - H}{H_0} \right) \times 100\% \quad (1)$$

The wear resistance of ceramics before and after irradiation was determined by calculating changes in the dry friction coefficient using the tribological method. The number of test cycles was 20,000, and the load on the metal ball was 200 N. Based on the changes in the dry friction coefficient before and after 20,000 test cycles, the value of the coefficient deterioration was determined, which characterizes the loss of the wear resistance of the initial and irradiated materials to mechanical stress.

The determination of the effect of irradiation and subsequent radiation-induced defects accumulation and implanted helium concentration on the heat-conducting properties and a decrease in thermal conductivity was carried out using the standard method for determination of the longitudinal heat flux. This method was implemented using the KIT-800 device (Granat, Moscow, Russia).

3. Results and Discussion

Figure 2 shows the X-ray phase analysis data reflecting changes in the structural parameters of the samples under study depending on the irradiation fluence. According to the data of X-ray phase analysis, it was found that the samples under study have a hexagonal structure with the spatial system P63mc(186) and crystal lattice parameters a = 2.66986 Å, c = 4.33690 Å. Analysis of the obtained diffraction patterns showed that in the case of the initial samples, the shape of the diffraction lines, as well as the ratio of the areas of reflections and background radiation, indicate a high degree of ordering of the crystal structure (more than 98%). In this case, for irradiated samples, the main changes in the diffraction patterns shown in Figure 2 are associated with two types of changes.

The first type is associated with a change in the intensities and shape of diffraction peaks, caused by the processes of crushing and recrystallization of grains under the action of irradiation, as well as a change in their orientation [35,36]. The appearance of the effect of reorientation of grains as a result of their mobility under the action of irradiation is evidenced by the fact that, at doses of 10^{15}–10^{17} ion/cm^2, a change in the intensity of the (100), (002), and (101) reflections is observed, with a clearly observed increase in the intensity of the (002) and (101) reflections. This behavior of changes in the intensities of reflections indicates a reorientation of grains under the action of irradiation, caused by the transfer of the kinetic energy of incident ions into the structure of the irradiated layer, followed by the transformation of kinetic energy into thermal energy. This transformation leads to an increase in the thermal vibrations of atoms in the lattice, as well as local heating of the structure. As a result of such influences, a partial reorientation of crystallites occurs, due to both the processes of mobility and the processes of crushing and subsequent amorphization. It should be mentioned that in the work with similar types of commercial ceramics exposed to helium irradiation, it was assumed that the change in the shape of the (002) reflections at high irradiation doses may be associated with polymorphic transformations of the BeO-hexagonal → BeO-cubic type [36]. Such polymorphic transformations can be caused by the crystal structure disordering and partial amorphization processes, which lead to the formation of impurity inclusions of the cubic phase at high radiation doses. In this case, a detailed analysis of the shape of the diffraction reflection (002) at irradiation doses above 10^{17} ion/cm^2 revealed a strong asymmetry of the reflection characteristic of the formation of impurity inclusions of the cubic BeO phase, the content of which is no more than 3–5%.

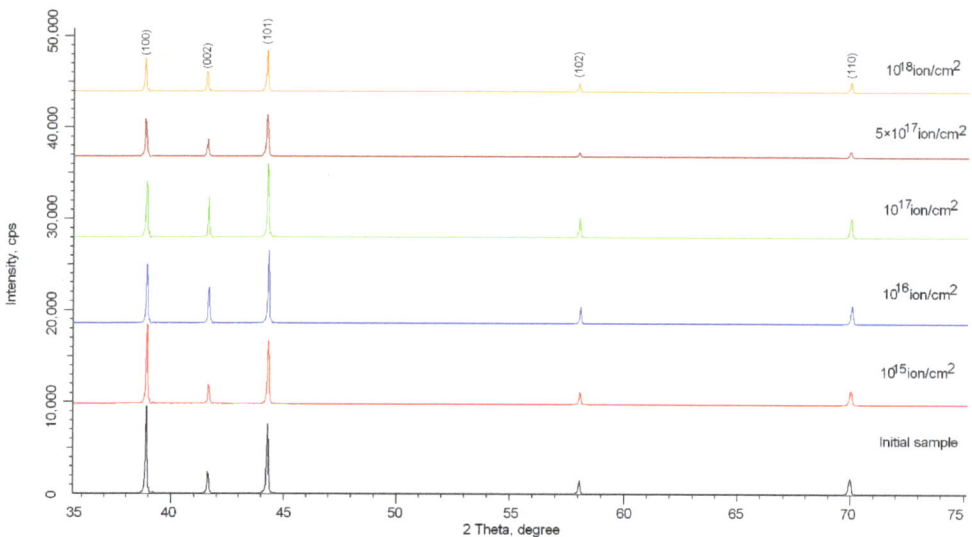

Figure 2. X-ray diffraction patterns of the studied ceramic samples versus irradiation fluence.

The second type of changes is caused by the shift of diffraction maxima to the region of small angles, which indicates crystal lattice deformation and swelling processes under the action of irradiation. A detailed representation of the change in shape and position of the main diffraction lines (100), (002), and (101), reflecting the change in the crystal lattice, is shown in Figure 3a. As can be seen from the presented data, the greatest shift of diffraction reflections is observed for fluences of 10^{16}–10^{17} ion/cm^2, while the change in intensities for these fluences is associated only with crystallites reorientation processes and a change in texture. A further increase in the irradiation fluence to 5×10^{17}–10^{18} ion/cm^2 leads to a sharp decrease in the intensity of reflections, as well as an increase in the asymmetry of reflections, which, as mentioned above, can be caused by the formation of inclusions of a cubic phase in ceramic structure [37,38]. At the same time, the change in diffraction maxima positions for these irradiation fluences is less pronounced than for lower fluences (see Figure 3b). This behavior may be due to the fact that at these fluences, the dominant radiation damage mechanism is amorphization and the formation of impurity phases in the structure.

Table 1 shows the results of changes in the crystal lattice parameters of the studied ceramics depending on the irradiation fluence. As can be seen from the presented data, an increase in the irradiation fluence leads to a shift in the position of diffraction reflections, and, consequently, a distortion of the crystal lattice leads to an increase in the lattice parameters, as well as its volume. Increase in the crystal lattice volume as a result of irradiation indicates crystal structure swelling due to both deformation and helium ion implantation processes, followed by the formation of gas-filled bubbles in the structure.

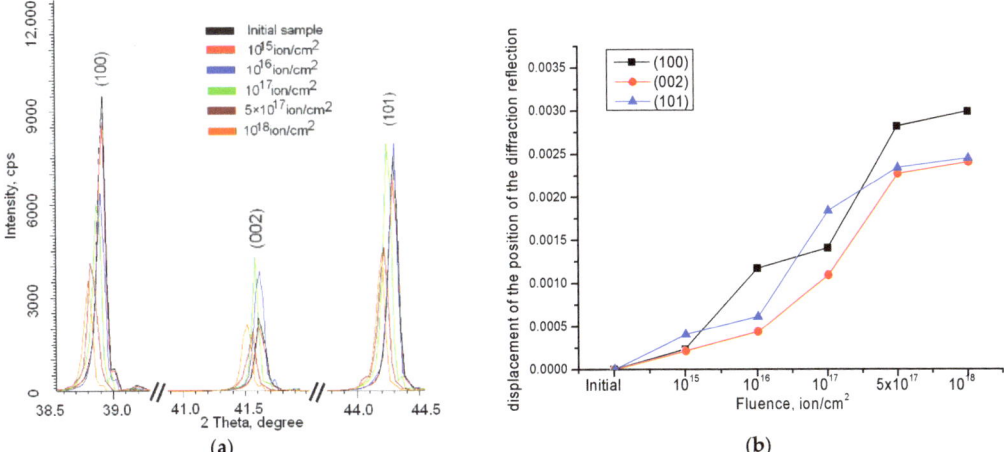

Figure 3. (**a**) Dynamics of changes in the main diffraction reflections depending on irradiation fluence. (**b**) Results of change in the shift of diffraction reflections (110), (002), and (101).

Table 1. Crystal lattice data.

Fluence, ion/cm^2	Initial Sample	10^{15}	10^{16}	10^{17}	5×10^{17}	10^{18}
Lattice parameter, Å	a = 2.66986, c = 4.33690	a = 2.67039, c = 4.33730	a = 2.67100, c = 4.33966	a = 2.67312, c = 4.34139	a = 2.67678, c = 4.34485	a = 2.67785, c = 4.34658
c/a	1.6243	1.6242	1.6247	1.6241	1.6232	1.6231
Lattice volume, Å^3	26.77	26.79	26.81	26.87	26.96	26.99

Figure 4 shows the results of crystal lattice swelling determined according to Equation (2):

$$Swelling = \left(\frac{V - V_0}{V_0}\right) \times 100\% \qquad (2)$$

where V and V_0 are values of the crystal lattice volume for irradiated and initial samples.

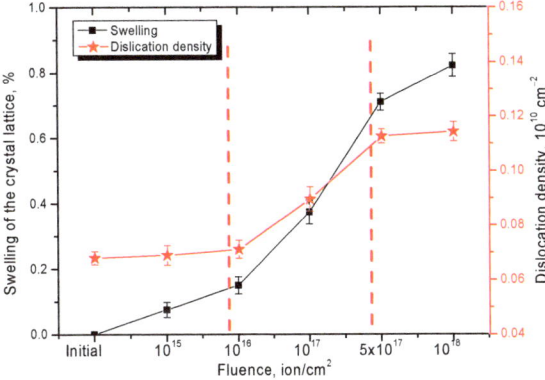

Figure 4. Swelling of the crystal lattice and changes in the dislocation density in the ceramic structure.

The dislocation density in our case was estimated by the standard method based on a change in the size of crystallites, which are estimated from X-ray data [36,37]. The following Equation (3) is used as a calculation formula:

$$Dislocation\_density = \frac{1}{D^2}, \qquad (3)$$

where D is the crystallite size determined from the analysis of X-ray diffraction patterns.

As can be seen from the data presented, the change in crystal lattice volume and, consequently, its swelling has a three-stage nature, characterized by different trends in the increase in swelling. At fluences of 10^{15}–10^{16} ion/cm^2, the swelling of the crystal lattice is insignificant, which is due to the fact that at these fluences, the main processes caused by irradiation are the processes of reorientation of crystallites and deformation of the structure due to the agglomeration of defects. At the same time, at the given irradiation fluences, changes in dislocation density are also insignificant.

The second stage of swelling changes is associated with a sharp change in the swelling trend and an increase in swelling from 0.15% to 0.7%. The swelling of the crystal lattice at these fluences is primarily associated not only with deformation processes, but also with the accumulation of implanted helium, which leads to the formation of defect agglomerates. In this case, a 1.5-fold increase in the dislocation density is also observed, which indicates a decrease in the grain size as a result of crushing and amorphization.

The third stage of changes is typical for fluences 5×10^{17}–10^{18} ion/cm^2 and is characterized by small changes in the swelling and dislocation density, which indicates a decrease in the rate of defect accumulation in the structure and the so-called saturation effect. Moreover, a decrease in the swelling rate can be due to the formation of impurity inclusions in the structure of a cubic phase, leading to amorphization of the structure.

Some of the important performance characteristics of ceramics are their mechanical and strength properties, as well as the dynamics of their change during irradiation and operation. It is a known fact that, under irradiation with heavy ions with low energies, a hardening of the surface layer is observed [42–44]. This is primarily due to the processes of change in the dislocation density, leading to radiation hardening. However, this effect has a strong dose dependence, and is observed mainly for irradiation doses of 10^{12}–10^{15} ions/cm^2, which are characterized by the formation of dislocation and point defects, leading to the formation of a strengthening layer. In our case, the irradiation was carried out with He^{2+} ions, which by their nature have high mobility and low solubility in the structure, leading to the formation of agglomerates in the structure in the form of gas-filled regions and bubbles. At the same time, in our case, the irradiation doses were 10^{15}–10^{18} ion/cm^2, which is much higher than the doses typical for observing the hardening effect.

Figure 5 shows the results of changes in the hardness of the near-surface layer of ceramics as a result of the accumulation of radiation-induced damage and implanted ions. With an irradiation fluence of 10^{15}–10^{16} ion/cm^2, the change in the hardness indicators is insignificant, and is no more than 1–3%, which indicates a high resistance of materials to radiation-induced damage caused by irradiation. With these fluences, damage accumulates in the structure of the surface layer due to the formation of dislocation and cluster defects, as well as the possible agglomeration of implanted helium into gas-filled bubbles. An increase in the radiation dose to 10^{17}–5×10^{17} ion/cm^2 leads to a sharp decrease in hardness indicators and an increase in the softening degree of the near-surface layer from 3% to 13–32%, which is a 5–10-fold decrease in the degree of resistance to softening and embrittlement. This destructive behavior of changes in strength properties is associated with a sharp crystal structure swelling, leading to the rupture of chemical and crystal bonds. The increase in swelling is due to a rise in the implanted helium concentration, which leads to an increase in the volumes of gas-filled bubbles and, consequently, to an increase in internal pressure on the crystal structure. A pressure increase in the crystal structure leads to an increase in deformation and distortions of the crystal lattice, which is also associated with an increase in atomic displacements, leading to disorientation of the crystal structure and its amorphization. As is known from the literature, radiation doses above 10^{17} ion/cm^2 are critical for oxide and nitride ceramics, as well as multilayer radiation-resistant coatings, which are associated with the formation of blister inclusions and gas-filled bubbles of various diameters in the structure of the irradiated near-surface layer. The formation of such defects leads to embrittlement and partial destruction of the near-surface layer, which entails a decrease in mechanical and strength characteristics. It is known that the processes of the accumulation of radiation damage in the structure of the surface layer are nonlinear, and at certain doses, a decrease in the degree of radiation damage is observed, which is due to the effect of defect accumulation in the structure and amorphization processes [38,45]. This behavior for selected ceramics is observed at a dose above 5×10^{17} ion/cm^2, which consists of a sharp change in the trend of the decrease of irradiated ceramic strength properties, as well as the softening and embrittlement of the near-surface layer. This is primarily due to an increase in the degree of the amorphization of the irradiated layer and its swelling due to the accumulation of implanted helium in the structure.

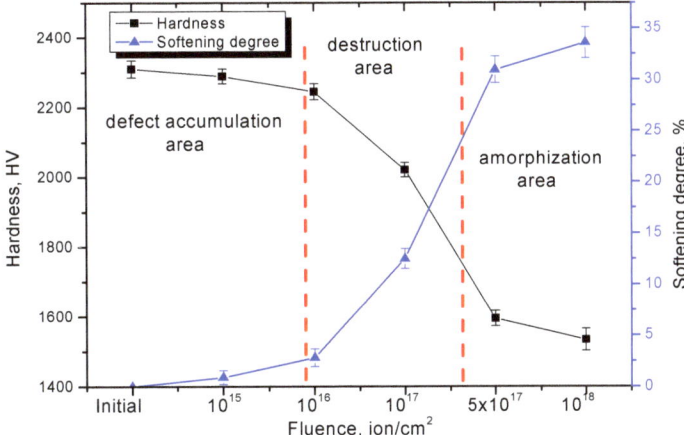

Figure 5. Results of changes in hardness and softening of the ceramic near-surface layer depending on irradiation fluence.

Figure 6 shows the results of the changes in the dry friction coefficient, reflecting the wear resistance of ceramics to external mechanical influences. As can be seen from the data presented, changes in the dry friction coefficient can be divided into two factors, reflecting both wear resistance over a long number of cycles and the change in surface defectiveness as a result of irradiation. At irradiation fluences of 10^{15}–10^{17} ion/cm^2, change in the dry friction coefficient is insignificant, which indicates a small degree of surface defectiveness as a result of irradiation and subsequent deformation processes caused by the accumulation of radiation-induced defects in the surface layer structure. The main changes in the dry friction coefficient for these irradiation fluences are observed after 15,000 test cycles, when the coefficient increases by 15–25%, which indicates a surface deterioration and a decrease in wear resistance.

For fluences of 5×10^{17}–10^{18} ion/cm^2, an increase in the dry friction coefficient is observed from 0.34 (initial sample) to 0.42 and 0.51, respectively, which indicates a surface deterioration and the formation of additional defects or irregularities leading to an increase in friction resistance. This behavior may be due to the defect accumulation in the structure, as well as partial amorphization, which leads to a sharp deterioration of not only hardness, but also wear resistance. Furthermore, a decrease in resistance to mechanical stress is evidenced by a sharp deterioration in the dry friction coefficient after 15,000 tests, which indicates the low stability of the near-surface irradiated layer to mechanical stress.

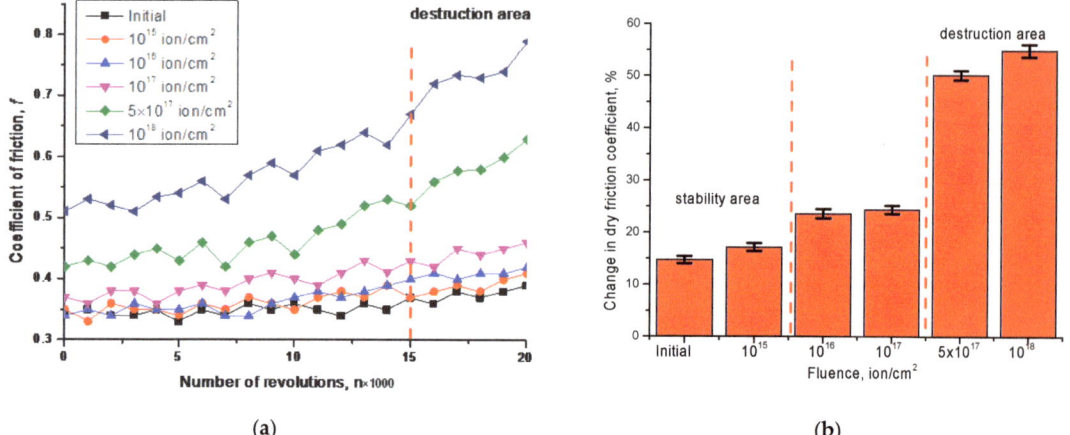

Figure 6. (a) Results of changes in the dry friction coefficient depending on the number of test cycles; (b) Results of dry friction coefficient deterioration depending on irradiation fluence.

Figure 7 shows the results of changes in the heat-conducting properties of ceramics, as well as thermal conductivity loss depending on irradiation fluence and accumulated radiation damage concentration.

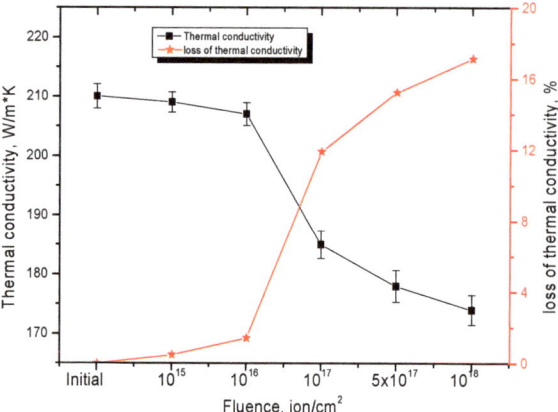

Figure 7. Results of changes in the heat-conducting properties of ceramics before and after irradiation.

The general appearance of the trend in the thermal conductivity coefficient is similar to changes in the strength and structural properties of ceramics, which indicates a direct dependence of the effect of radiation-induced defects concentration in the irradiated near-surface layer structure on the heat-conducting properties. In this case, accumulation of radiation-induced damage, entailing deformation and swelling of the crystal structure, leads to a decrease in heat-conducting properties. However, the formation of impurity inclusions in the structure of the irradiated layer slows down thermal conductivity deterioration. Thus, the obtained dependences of changes in heat-conducting properties indicate that at irradiation doses of 10^{15}–10^{16} ions/cm^2, heat losses are no more than 1–2%, which is within acceptable limits. However, the accumulation of radiation-induced defects, as well as the concentration of implanted helium in the structure of the surface layer, leads to sharp thermal conductivity deterioration and significant heat losses exceeding 10% of the initial value.

4. Conclusions

In conclusion, we can summarize the results of the studies carried out aimed at obtaining new information on the radiation resistance of BeO ceramics to the helium swelling processes and subsequent near-surface layer destruction. During research, it was found that the change in structural properties depending on irradiation fluence occurs in two stages associated with different mechanisms of radiation-induced defect accumulation. It was found that at irradiation fluences above 10^{17} ion/cm^2, the dominant radiation damage mechanism is amorphization and formation of impurity phases in the structure. In turn, the rate and value of the crystal lattice swelling is directly dependent on implanted helium concentration and subsequent partial amorphization and disordering processes of the crystal structure. During the study of changes in the mechanical and strength properties of ceramics depending on irradiation fluence, it was found that the destructive behavior of changes in strength properties is directly related to the crystal structure swelling, as well as the radiation-induced defect accumulation rate. During the study of heat-conducting properties, it was found that with an irradiation fluence of 10^{15}–10^{16} ion/cm^2, the decrease in thermal conductivity is minimal and within the measurement error, while an increase in irradiation fluence above 10^{17} ion/cm^2 leads to an increase in heat loss by more than 10%.

Author Contributions: Conceptualization, M.V.Z., D.I.S., D.B.B. and A.L.K.; methodology, D.B.B., D.I.S. and A.L.K.; formal analysis, M.V.Z.; investigation, D.I.S., A.L.K. and M.V.Z.; resources, M.V.Z.; writing—original draft preparation, review and editing, D.I.S., M.V.Z., D.B.B. and A.L.K.; visualization, M.V.Z.; supervision, M.V.Z. All authors have read and agreed to the published version of the manuscript.

Funding: This research was funded by the Science Committee of the Ministry of Education and Science of the Republic of Kazakhstan (No. AP08855828).

Institutional Review Board Statement: Not applicable.

Informed Consent Statement: Not applicable.

Data Availability Statement: Not applicable.

Conflicts of Interest: The authors declare that they have no conflict of interest.

References

1. Neeft, E.; Bakker, K.; Schram, R.; Conrad, R.; Konings, R. The EFTTRA-T3 irradiation experiment on inert matrix fuels. *J. Nucl. Mater.* **2003**, *320*, 106–116. [CrossRef]
2. Gong, X.; Ding, S.; Zhao, Y.; Huo, Y.; Zhang, L.; Li, Y. Effects of irradiation hardening and creep on the thermo-mechanical behaviors in inert matrix fuel elements. *Mech. Mater.* **2013**, *65*, 110–123. [CrossRef]
3. Zhang, J.; Wang, H.; Wei, H.; Tang, C.; Lu, C.; Huang, C.; Ding, S.; Li, Y. Modelling of effective irradiation swelling for inert matrix fuels. *Nucl. Eng. Technol.* **2021**, *53*, 2616–2628. [CrossRef]
4. Korneeva, E.; Ibrayeva, A.; van Vuuren, A.J.; Kurpaska, L.; Clozel, M.; Mulewska, K.; Kirilkin, N.; Skuratov, V.; Neethling, J.; Zdorovets, M. Nanoindentation testing of Si3N4 irradiated with swift heavy ions. *J. Nucl. Mater.* **2021**, *555*, 153120. [CrossRef]
5. Luzzi, L.; Barani, T.; Boer, B.; Cognini, L.; Del Nevo, A.; Lainet, M.; Lemehov, S.; Magni, A.; Marelle, V.; Michel, B.; et al. Assessment of three European fuel performance codes against the SUPERFACT-1 fast reactor irradiation experiment. *Nucl. Eng. Technol.* **2021**, *53*, 3367–3378. [CrossRef]
6. Xiang, F.; He, Y.; Niu, Y.; Deng, C.; Wu, Y.; Wang, K.; Tian, W.; Su, G.; Qiu, S. A new method to simulate dispersion plate-type fuel assembly in a multi-physics coupled way. *Ann. Nucl. Energy* **2022**, *166*, 108734. [CrossRef]
7. Boer, B.; Lemehov, S.; Wéber, M.; Parthoens, Y.; Gysemans, M.; McGinley, J.; Somers, J.; Verwerft, M. Irradiation performance of (Th,Pu)O$_2$ fuel under Pressurized Water Reactor conditions. *J. Nucl. Mater.* **2016**, *471*, 97–109. [CrossRef]
8. Sarma, K.; Fourcade, J.; Lee, S.-G.; Solomon, A. New processing methods to produce silicon carbide and beryllium oxide inert matrix and enhanced thermal conductivity oxide fuels. *J. Nucl. Mater.* **2006**, *352*, 324–333. [CrossRef]
9. Klaassen, F.; Bakker, K.; Schram, R.; Meulekamp, R.K.; Conrad, R.; Somers, J.; Konings, R. Post irradiation examination of irradiated americium oxide and uranium dioxide in magnesium aluminate spinel. *J. Nucl. Mater.* **2003**, *319*, 108–117. [CrossRef]
10. Lamontagne, J.; Béjaoui, S.; Hanifi, K.; Valot, C.; Loubet, L.; Valot, C. Swelling under irradiation of MgO pellets containing americium oxide: The ECRIX-H irradiation experiment. *J. Nucl. Mater.* **2011**, *413*, 137–144. [CrossRef]
11. Revankar, S.T.; Zhou, W.; Chandramouli, D. Thermal Performance of UO$_2$-BeO Fuel during a Loss of Coolant Accident. *Int. J. Nucl. Energy Sci. Eng.* **2015**, *5*, 1. [CrossRef]

12. Shelley, A. Neutronic analyses of americium burning U-free inert matrix fuels. *Prog. Nucl. Energy* **2020**, *130*, 103567. [CrossRef]
13. Zhu, X.; Gao, R.; Gong, H.; Liu, T.; Lin, D.Y.; Song, H. UO_2/BeO interfacial thermal resistance and its effect on fuel thermal conductivity. *Ann. Nucl. Energy* **2021**, *154*, 108102. [CrossRef]
14. Kiiko, V.S.; Vaispapir, V.Y. Thermal Conductivity and Prospects for Application of BeO Ceramic in Electronics. *Glas. Ceram.* **2015**, *71*, 387–391. [CrossRef]
15. Akishin, G.P.; Turnaev, S.K.; Vaispapir, V.Y.; Gorbunova, M.A.; Makurin, Y.N.; Kiiko, V.S.; Ivanovskii, A.L. Thermal conductivity of beryllium oxide ceramic. *Refract. Ind. Ceram.* **2009**, *50*, 465–468. [CrossRef]
16. Jahn, A.; Sommer, M.; Ullrich, W.; Wickert, M.; Henniger, J. The BeOmax system—Dosimetry using OSL of BeO for several applications. *Radiat. Meas.* **2013**, *56*, 324–327. [CrossRef]
17. Bulur, E.; Saraç, B. Time-resolved OSL studies on BeO ceramics. *Radiat. Meas.* **2013**, *59*, 129–138. [CrossRef]
18. Altunal, V.; Guckan, V.; Yu, Y.; Dicker, A.; Yegingil, Z. A newly developed OSL dosimeter based on beryllium oxide: BeO: Na, Dy, Er. *J. Lumin.* **2020**, *222*, 117140. [CrossRef]
19. Yukihara, E. A review on the OSL of BeO in light of recent discoveries: The missing piece of the puzzle? *Radiat. Meas.* **2020**, *134*, 106291. [CrossRef]
20. Altunal, V.; Guckan, V.; Ozdemir, A.; Yegingil, Z. Radiation dosimeter utilizing optically stimulated luminescence of BeO: Na, Tb Gd ceramics. *J. Alloys Compd.* **2020**, *817*, 152809. [CrossRef]
21. Altunal, V.; Guckan, V.; Ozdemir, A.; Can, N.; Yegingil, Z. Luminescence characteristics of Al-and Ca-doped BeO obtained via a sol-gel method. *J. Phys. Chem. Solids* **2019**, *131*, 230–242. [CrossRef]
22. Wójcik, N.A.; Tagiara, N.S.; Ali, S.; Górnicka, K.; Segawa, H.; Klimczuk, T.; Kamitsos, E.I. Structure and magnetic properties of BeO-Fe_2O_3-Al_2O_3-TeO_2 glass-ceramic composites. *J. Eur. Ceram. Soc.* **2021**, *41*, 5214–5222. [CrossRef]
23. Aşlar, E.; Meric, N.; Şahiner, E.; Erdem, O.; Kitis, G.; Polymeris, G.S. A correlation study on the TL, OSL and ESR signals in commercial BeO dosimeters yielding intense transfer effects. *J. Lumin.* **2019**, *214*, 116533. [CrossRef]
24. Altunal, V.; Guckan, V.; Ozdemir, A.; Ekicibil, A.; Karadag, F.; Yegingil, I.; Zydhachevskyy, Y. A systematic study on luminescence characterization of lanthanide-doped BeO ceramic dosimeters. *J. Alloys Compd.* **2021**, *876*, 160105. [CrossRef]
25. Trukhanov, A.V.; Kozlovskiy, A.L.; Ryskulov, A.E.; Uglov, V.V.; Kislitsin, S.B.; Zdorovets, M.V.; Trukhanov, S.V.; Zubar, T.I.; Astapovich, K.A.; Tishkevich, D.I. Control of structural parameters and thermal conductivity of BeO ceramics using heavy ion irradiation and post-radiation annealing. *Ceram. Int.* **2019**, *45*, 15412–15416. [CrossRef]
26. Snead, L.; Zinkle, S.; White, D. Thermal conductivity degradation of ceramic materials due to low temperature, low dose neutron irradiation. *J. Nucl. Mater.* **2005**, *340*, 187–202. [CrossRef]
27. Ryazanov, A.I.; Klaptsov, A.V.; Kohyama, A.; Katoh, Y.; Kishimoto, H. Effect of Helium on Radiation Swelling of SiC. *Phys. Scr.* **2004**, *T111*, 195. [CrossRef]
28. Ghoniem, N.; Takata, M. A rate theory of swelling induced by helium and displacement damage in fusion reactor structural materials. *J. Nucl. Mater.* **1982**, *105*, 276–292. [CrossRef]
29. Wang, H.; Qi, J.; Guo, H.; Chen, R.; Yang, M.; Gong, Y.; Huang, Z.; Shi, Q.; Liu, W.; Wang, H.; et al. Influence of helium ion radiation on the nano-grained Li_2TiO_3 ceramic for tritium breeding. *Ceram. Int.* **2021**, *47*, 28357–28366. [CrossRef]
30. Scaffidi-Argentina, F.; Donne, M.; Ferrero, C.; Ronchi, C. Helium induced swelling and tritium trapping mechanisms in irradiated beryllium: A comprehensive approach. *Fusion Eng. Des.* **1995**, *27*, 275–282. [CrossRef]
31. Allen, W.; Zinkle, S. Lattice location and clustering of helium in ceramic oxides. *J. Nucl. Mater.* **1992**, *191–194*, 625–629. [CrossRef]
32. Van Veen, A.; Konings, R.; Fedorov, A. Helium in inert matrix dispersion fuels. *J. Nucl. Mater.* **2003**, *320*, 77–84. [CrossRef]
33. Su, Q.; Inoue, S.; Ishimaru, M.; Gigax, J.; Wang, T.; Ding, H.; Demkowicz, M.J.; Shao, L.; Nastasi, M. Helium irradiation and implantation effects on the structure of amorphous silicon oxycarbide. *Sci. Rep.* **2017**, *7*, 3900. [CrossRef]
34. Liu, P.; Xue, L.; Yu, L.; Liu, J.; Hu, W.; Zhan, Q.; Wan, F. Microstructure change and swelling of helium irradiated beryllium. *Fusion Eng. Des.* **2019**, *140*, 62–66. [CrossRef]
35. Shi, K.; Shu, X.; Shao, D.; Zhang, H.; Liu, Y.; Peng, G.; Lu, X. Helium ions' irradiation effects on $Gd_2Zr_2O_7$ ceramics holding complex simulated radionuclides. *J. Radioanal. Nucl. Chem.* **2017**, *314*, 2113–2122. [CrossRef]
36. Kislitsin, S.; Ryskulov, A.; Kozlovskiy, A.; Ivanov, I.; Uglov, V.; Zdorovets, M. Degradation processes and helium swelling in beryllium oxide. *Surf. Coat. Technol.* **2020**, *386*, 125498. [CrossRef]
37. Kozlovskiy, A.L.; Zdorovets, M.V. Study of the radiation disordering mechanisms of AlN ceramic structure as a result of helium swelling. *J. Mater. Sci. Mater. Electron.* **2021**, *32*, 21658–21669. [CrossRef]
38. Shlimas, D.I.; Zdorovets, M.V.; Kozlovskiy, A.L. Synthesis and resistance to helium swelling of Li_2TiO_3 ceramics. *J. Mater. Sci. Mater. Electron.* **2020**, *31*, 12903–12912. [CrossRef]
39. Matsunaga, J.; Sakamoto, K.; Muta, H.; Yamanaka, S. Dependence of vacancy concentration on morphology of helium bubbles in oxide ceramics. *J. Nucl. Sci. Technol.* **2014**, *51*, 1231–1240. [CrossRef]
40. Ziegler, J.F.; Ziegler, M.D.; Biersack, J.P. SRIM–The stopping and range of ions in matter. *Nucl. Instrum. Methods Phys. Res. Sect. B Beam Interact. Mater. At.* **2010**, *268*, 1818–1823. [CrossRef]

41. Egeland, G.; Valdez, J.; Maloy, S.; McClellan, K.; Sickafus, K.; Bond, G. Heavy-ion irradiation defect accumulation in ZrN characterized by TEM, GIXRD, nanoindentation, and helium desorption. *J. Nucl. Mater.* **2013**, *435*, 77–87. [CrossRef]
42. Kadyrzhanov, K.K.; Tinishbaeva, K.; Uglov, V.V. Investigation of the effect of exposure to heavy Xe22+ ions on the mechanical properties of carbide ceramics. *Eurasian Phys. Tech. J.* **2020**, *17*, 46–53. [CrossRef]
43. Tinishbaeva, K.; Kadyrzhanov, K.K.; Kozlovskiy, A.L.; Uglov, V.V.; Zdorovets, M.V. Implantation of low-energy Ni^{12+} ions to change structural and strength characteristics of ceramics based on SiC. *J. Mater. Sci. Mater. Electron.* **2019**, *31*, 2246–2256. [CrossRef]
44. Jin, K.; Lu, C.; Wang, L.; Qu, J.; Weber, W.; Zhang, Y.; Bei, H. Effects of compositional complexity on the ion-irradiation induced swelling and hardening in Ni-containing equiatomic alloys. *Scr. Mater.* **2016**, *119*, 65–70. [CrossRef]
45. Li, N.; Fu, E.; Wang, H.; Carter, J.; Shao, L.; Maloy, S.; Misra, A.; Zhang, X. He ion irradiation damage in Fe/W nanolayer films. *J. Nucl. Mater.* **2009**, *389*, 233–238. [CrossRef]

Article

As-Sintered Manganese-Stabilized Zirconia Ceramics with Excellent Electrical Conductivity

Ling Gao [1,*], Ruidong Guan [1], Shengnan Zhang [2], Hao Zhi [3], Changqing Jin [1], Lihua Jin [2], Yongxing Wei [1] and Jianping Wang [4,*]

[1] School of Materials Science and Chemical Engineering, Xi'an Technological University, Xi'an 710021, China; guanruidong1998@163.com (R.G.); jinchangqing@xatu.edu.cn (C.J.); weiyongxing@xatu.edu.cn (Y.W.)
[2] Northwest Institute for Nonferrous Metal Research, Xi'an 710016, China; snzhang@c-nin.com (S.Z.); lhjin@c-nin.com (L.J.)
[3] Xi'an Sailong Metal Materials Co., Ltd., Export Processing Zone, No.1 Ming Guang Road, Xi'an 710018, China; zhihao1983@163.com
[4] School of Materials Science and Engineering, YingKou Institute of Technology, Yingkou 115014, China
* Correspondence: gaoling@xatu.edu.cn (L.G.); 0wangjianping@163.com (J.W.)

Abstract: As promising electrolyte materials in intermediate-temperature solid oxide fuel cells (IT-SOFCs), Sc-stabilized ZrO_2 (ScSZ) and Y-stabilized ZrO_2 (YSZ) electrolytes continue to be plagued by high cost and low intermediate conductivity. To mitigate these problems, Mn has been chosen as a new stabilization element for the synthesis of Mn-stabilized ZrO_2 ceramics (MnSZ) through solid state sintering. Microstructures and electrical properties of micron-crystalline $Zr_{1-x}Mn_xO_{2-\delta}$ ($x = 0.15$, 0.20 and 0.25) ceramics electrolytes for IT-SOFCs have been systematically evaluated. Within the applied doping content, Mn^{2+} ions can enter the ZrO_2 crystal lattice, leading to the formation of single cubic phase samples. Electrical conductivity measurements in the temperature range between 400 °C and 1000 °C show a sharp increase in conductivity due to Mn doping. The highest conductivity is obtained from the $Zr_{0.75}Mn_{0.25}O_{2-\delta}$ samples, being 0.0144 S/cm at 600 °C and 0.182 S/cm at 1000 °C. The electrical conductivity at 600 °C is twice higher than that of the YSZ and two orders of magnitude higher than that of the ScSZ. These properties can fulfill the conductivity requirement ($\sim 1 \times 10^{-2}$ S/cm) for the electrolyte. Therefore, based on this study, we propose that Mn stabilized ZrO_2 is a promising candidate as a solid electrolyte for IT-SOFCs.

Keywords: intermediate-temperature solid oxide fuel sells; Mn-stabilized zirconia ceramics; microstructure; electrical property

Citation: Gao, L.; Guan, R.; Zhang, S.; Zhi, H.; Jin, C.; Jin, L.; Wei, Y.; Wang, J. As-Sintered Manganese-Stabilized Zirconia Ceramics with Excellent Electrical Conductivity. *Crystals* **2022**, *12*, 620. https://doi.org/10.3390/cryst12050620

Academic Editors: Linghang Wang and Gang Xu

Received: 12 March 2022
Accepted: 6 April 2022
Published: 27 April 2022

Publisher's Note: MDPI stays neutral with regard to jurisdictional claims in published maps and institutional affiliations.

Copyright: © 2022 by the authors. Licensee MDPI, Basel, Switzerland. This article is an open access article distributed under the terms and conditions of the Creative Commons Attribution (CC BY) license (https://creativecommons.org/licenses/by/4.0/).

1. Introduction

Solid oxide fuel cells (SOFC) have been attracting attention due to their efficient conversion of electrochemical fuel to electricity with negligible pollution. From a view of long-term durability and reliability, the current developmental target for SOFCs is to reduce the operating temperature into the intermediate temperature (IT) range (500–700 °C), which requires increased electrolyte ionic conductivity and enhanced gas/electrode reaction kinetics [1]. YSZ electrolyte has been recognized to be the most promising commercial applications candidate for SOFCs. However, the relative lower ionic conductivity of YSZ at intermediate temperature limits its application as an electrolyte candidate for IT-SOFCs [2]. The ScSZ electrolyte exhibits the highest ionic conductivity among all zirconia-based materials, and is considered to be the best choice for IT-SOFCs electrolyte. However, the high costing and cubic-rhombohedral phase transformation around 600 °C make ScSZ inapplicable as an electrolyte [3–5]. Thus, an important practical task is the development of new zirconia-based electrolyte materials aimed at reducing the cost and increasing the intermediate temperature ionic conductivity.

In fact, ZrO_2 is a remarkable material, wherein specific cubic zirconia crystal structures can be stabilized by many kinds of dopants, which also enhances the electrical conductivity [6]. Though rare-earth metallic oxides have been used to make cubic phased zirconia-based electrolytes, studies show that transition metal oxide (especially MnO_x) can be an effective stabilizer on making cubic ZrO_2 [7,8]. In most previous studies, the phase composition, microstructure, catalytic property, and magnetic property of Mn-stabilized ZrO_2 (MnSZ) powders were systematically studied [9–13]. Many researchers have paid attention to the conductivity property of Mn-doped YSZ [14,15], and confirmed that Mn ion can considerably affect the conductivity of YSZ. Kim [16] found that with the addition of Mn_2O_3 in YSZ, the conductivity increase to 0.5 mS/cm at 600 °C. Lei [17] reported that 11ScSZ-2Mn_2O_3 ceramics with the cubic structure possesses the conductivity 4 mS/cm at 600 °C. Recently, researchers have studied the conductivity of 30 mol/% MnO doped ZrO_2 powders with the method of mechanical alloying and found that the intermediate temperature conductivity can be 0.4 mS/cm at 550 °C [18].

Mn oxide acts as a stabilizer, and has unique advantages: firstly, the reserves and costing of manganite is cheaper than yttria and scandia, which is beneficial for reducing battery manufacturing costs; secondly, the cubic phase structure of MnSZ is stable at a temperature lower than 800 °C at both pure O_2 and low oxygen partial pressure condition, which can solve the problem of phase transformation of ScSZ; thirdly, according to the defect reaction equation, with the same doping content, the oxygen vacancy number generated by charge compensation with divalent manganese ions replacement is more than trivalent yttrium and scandium, which means that MnSZ have higher electrical conductivity. So, we expect Mn-stabilized ZrO_2 to be an applicable IT-SOFC electrolyte.

In our previous study, we demonstrated that Mn doped ZrO_2 ceramics made by a co-precipitation method can be a potential electrode or electrolyte for SOFCs [19]. The highest conductivity is obtained from the $Zr_{0.8}Mn_{0.2}O_{2-\delta}$ samples being 0.002 S/cm at 600 °C. Due to the short sintering times and low doping content, the conductivity does not reach the requirement for electrolyte at 600 °C. However, the conductivity of 0.002 S/cm at 600 °C is three times higher than that of plasma-sprayed 8YSZ. It is noteworthy that the conductivity of ceramics largely depends on their phase composition, microstructure, density, and other parameters, which in turn depend on the synthesis technology. Solid state reaction method is a simple, cost effective, and mature technique to produce highly crystalline dense ceramics. Therefore, this method was used to process the binary composition with the minimum (15 at.%), intermediate (20 at.%), and maximum (25 at.%) doping content of cubic phase Mn-stabilized ZrO_2 [20]. The crystal structure, microstructure and the electro-conductivity were investigated for the as-synthesized bulk samples. The aim of this study is to produce cubic phased Mn-stabilized ZrO_2 ceramics and evaluate the electrical properties for IT-SOFCs applications.

2. Experiment

The samples were prepared by conventional solid state reaction in Ar. The starting materials for the study were zirconia (SCRC, 2–4 µm, purity: >99%) and manganese monoxide (Alfa Aesar, 2–4 µm, purity: >99.5%) powders. Both powders were weighted in the proper proportions and then homogeneously mixed in agate mortar. Then, the mixtures were uniaxially pressed into rectangular bars about $40 \times 5 \times 2$ mm^3 by applying a pressure of 10 MPa with the help of a hydraulic press. Finally, the green body was sintered in a tubular furnace at 1400 °C for 12 h in Ar with furnace cooling. The composition was designated as $Zr_{1-x}Mn_xO_{2-\delta}$ where x is the atomic percent of Mn, and its nominal content were 15, 20 and 25 at.%.

X-ray diffraction (Rigaku D/max-2550) with Cu-Kα radiation was used to analyze both the phases composition and the lattice parameters of the crashed powders. The XRD profiles were refined by the Rietveld method using the fullprof software and diffraction peak profiles were refined by a pseudo-Voigt function. The sintered samples were polished and thermal etched at 50 °C below the sintering temperature for 30 min, and the microstructure characterization was performed by SEM (JEOL JEM-6460). High resolution TEM and selected area electron diffraction (SAED) were performed with a CM200FEG microscope operating at 200 kV and equipped with a field-emission gun. X-ray photoelectron spectroscopy (XPS) was carried out on an ESCALAB 250Xi high-performance electron spectrometer (Thermo Fisher Company, Waltham, MA, USA), using monochromatic Al K-alpha radiation. Electrical conductivity was measured between 400 °C and 1000 °C in air by the 4-probe DC method. Platinum paste was applied to both current and voltage probes and the pasted samples were heated at 1000 °C for 1 h in air to cure the paste. The current source supplied the current in the −5 mA to +5 mA range with 1 mA step and the voltage drop was measured by a nano-voltmeter. The resistance values were calculated from the current-voltage curves. The electrical conductivity was measured while lowering temperature with 100 °C steps. The specimens were maintained for 1 h at each temperature before taking measurements.

3. Results and Discussion

Figure 1 shows the X-ray diffraction patterns along with their Rietveld refinement patterns of the as-sintered samples powders. It can be observed that in all the samples, a single cubic ZrO_2 phase was formed without other impurities and monoclinic phase. Because the solubility limit of Mn ions in ZrO_2 is reported to be 25 at.% [20], the chosen doping contents are all in the limit range. The almost linear residue of fittings of respective XRD patterns indicates a good agreement between experimental and calculated intensities. Following the Rietveld refinement calculation, the calculated lattice parameter of $Zr_{1-x}Mn_xO_{2-\delta}$ (x = 0.15, 0.2, 0.25) are 5.0889 Å, 5.0785 Å and 5.0771 Å respectively. A decrease in the lattice parameter was observed by increasing the Mn content, and the same result was also observed by other researchers [8]. According to the Vegard's law, substitution of Zr^{4+} (0.84 Å) by larger Mn^{2+} (0.96 Å) should result in the expansion of cubic lattice. The observed decrease of lattice parameters with Mn doping can be attributed to the creation of oxygen vacancy in c-ZrO_2 due to Mn doping. As the oxygen vacancy results in shrinkage of the cubic lattice, the resultant contraction of lattice parameter is dominated by the oxygen vacancy in the cubic lattice [18].

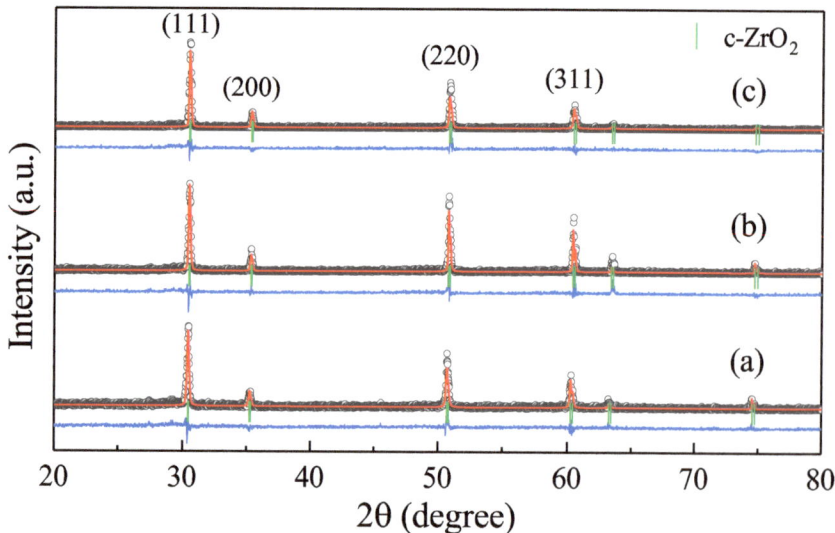

Figure 1. X-ray diffraction patterns along with their Rietveld refinement patterns of the $Zr_{1-x}Mn_xO_{2-\delta}$ (a) x = 0.15, (b) x = 0.20, (c) x = 0.25.

Figure 2a–c presents the SEM micrographs of the surface of $Zr_{1-x}Mn_xO_{2-\delta}$ specimens after thermal etching. Well-developed isometric grains are formed in all of the samples. Based on the statistics of a large number of grains, the average grain size of $Zr_{1-x}Mn_xO_{2-\delta}$

was determined to be 20, 50, and 60 μm, corresponding to the Mn contents of 0.15, 0.20, and 0.25, respectively. Transition-metal-oxides as sintering aids can facilitate the sintering process of zirconia [21]. It means that higher MnO doping content can result in increasing the grain growth of cubic zirconia. Due to the difference of the sintered precursor, the average grain sizes in this study are larger than those reported in our previous study [19]. However, inhomogeneous grain size distribution and residual porosity can also be observed in all the samples. Besides, compared to the clean grain boundary in the $x = 0.15$ and 0.20 samples, the impurities on the grain boundary of the $Zr_{0.75}Mn_{0.25}O_{2-\delta}$ sample can be obviously observed, which may be due to the grain boundary segregation of the doping element, though the solubility limit of Mn ions in ZrO_2 is reported to be 25 at.% [20]. To examine the intragrain microstructure, Figure 2d shows the typical HRTEM image of the $Zr_{0.8}Mn_{0.2}O_{2-\delta}$ sample. Lattice fringes observed in HRTEM appear regularly. The interplanar spacing value measured for (111) and (200) planes are 0.2936 nm and 0.2509 nm, respectively. Thus, this matrix can be assigned to the corresponding lattice structure of cubic ZrO_2. Selected area electron diffraction (SAED) pattern is obtained along [110] zone axis of the $Zr_{0.8}Mn_{0.2}O_{2-\delta}$ composition. Observed diffraction spots are identified and indexed with cubic ZrO_2 reflections. The HRTEM images and SAED patterns of the other two compositions are nearly the same with the $Zr_{0.8}Mn_{0.2}O_{2-\delta}$ samples. It signifies that all the samples in our study are entirely composed of c-ZrO_2 phase and there is no contamination.

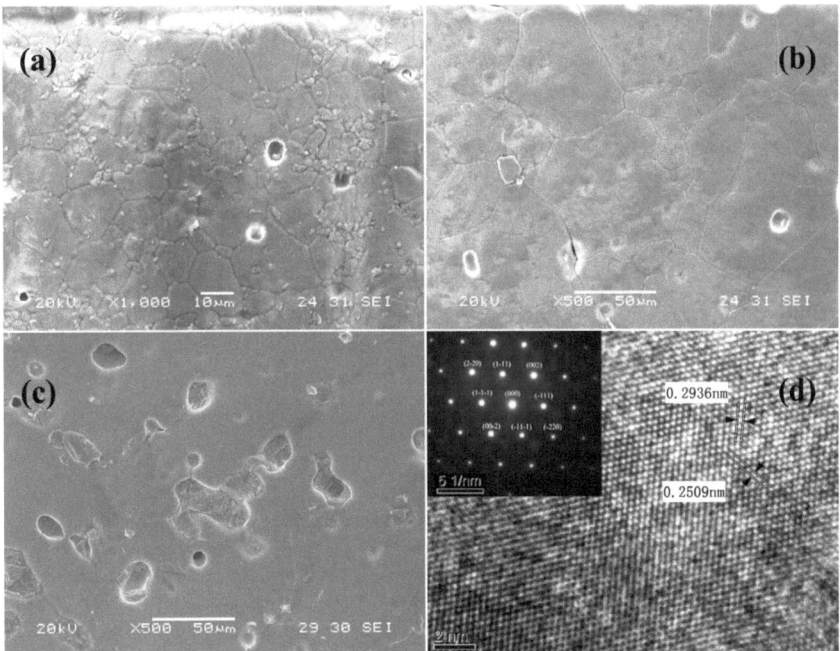

Figure 2. (**a–d**) SEM micrographs of $Zr_{1-x}Mn_xO_{2-\delta}$ solid solution (**a**) $x = 0.15$, (**b**) $x = 0.20$, (**c**) $x = 0.25$, HRTEM images and the SAED pattern from a single grain of $Zr_{0.80}Mn_{0.20}O_{2-\delta}$ along the [110] zone axis.

In order to explain why Mn ion doping can result in the formation of cubic ZrO_2, the chemical composition and elementary oxidation state of the powders crushed from $Zr_{1-x}Mn_xO_{2-\delta}$ ceramic were tested by XPS analysis. The high resolution spectra for Zr 3d, Mn 2p and O 1s are shown in Figure 3. The Zr 3d spectra of all the samples consist of Zr 3d $_{5/2}$ at 182.2 eV and Zr 3d $_{3/2}$ at 184.6 eV peaks with a peak separation of 2.4 eV, and correspond to the LS coupling value found for Zr (IV) in the oxide [22]. A closer comparison has found that the characterization peak of Zr 3d $_{5/2}$ shifts to higher binding energy by increasing the Mn content. Two main peaks from Mn $2p_{3/2}$ and Mn $2p_{1/2}$ can be observed. The Mn $2p_{3/2}$ peak is located at 641.6 eV with a satellite peak corresponding to the existence of Mn^{2+} in $Zr_{1-x}Mn_xO_{2-\delta}$ matrix. The XPS spectra of O 1s is used to investigate the presence of oxygen vacancies. Two separate peaks located at 530.3 eV and 531.7 eV can be attributed to the lattice oxygen and non-lattice oxygen (i.e. oxygen vacancy), respectively [23]. The relative amount of the oxygen vacancy are 32.31%, 32.9%, and 35.02% as obtained from the corresponding peak area. It indicates that the concentration of oxygen vacancy has increased with the increasing doping content. Cubic ZrO_2 containing higher MnO will yield higher ionic conduction. The XRD, SEM, and XPS results indicate that $Zr_{1-x}Mn_xO_{2-\delta}$ ceramics made with this method are a single cubic phased structure with a high concentration of oxygen vacancy. It means that $Zr_{1-x}Mn_xO_{2-\delta}$ ceramics can fulfill the structural requirement and can be a potential candidate for electrolyte.

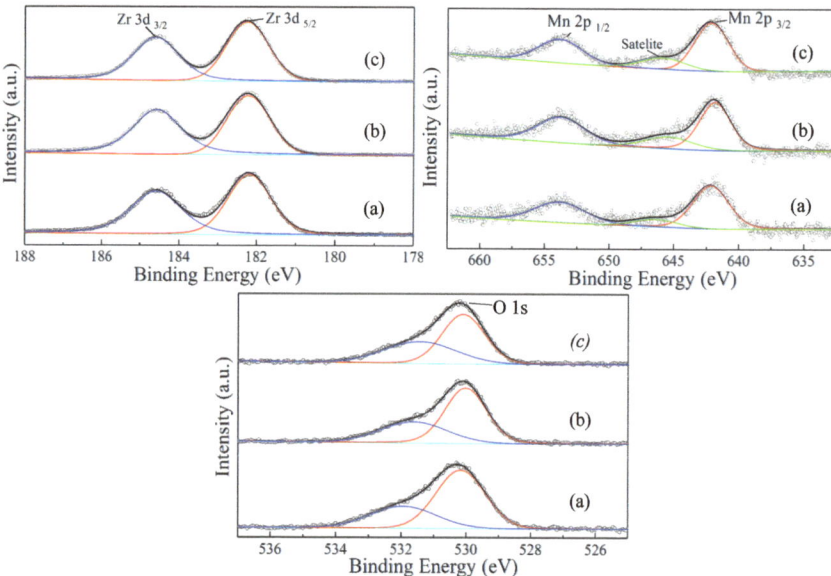

Figure 3. XPS high resolution spectrum of Zr 3d, Mn 2p and O 1s for $Zr_{1-x}Mn_xO_{2-\delta}$ (**a**) $x = 0.15$, (**b**) $x = 0.2$, (**c**) $x = 0.25$.

Figure 4 displays the typical Arrhenius plots of the total electrical conductivity of $Zr_{1-x}Mn_xO_{2-\delta}$ measured between 400 °C and 1000 °C in air with the method of 4-probe DC method. The total electrical conductivity increases with the increasing Mn content. As listed in Table 2, the single cubic phased $Zr_{1-x}Mn_xO_{2-\delta}$ samples exhibit high electro-conductivity at high temperatures. The highest conductivity reaches 0.0144 S/cm and 18.2 S/cm at 600 °C and 1000 °C, respectively. In a previous report, the highest ionic conductivity was obtained in 8 mol% YSZ electrolyte with the values of 0.0039 S/cm and 0.140 S/cm at 600 °C and 1000 °C, respectively [6]. Compared with these values, the conductivity of $Zr_{0.8}Mn_{0.2}O_{2-\delta}$ is nearly the same as 8YSZ at different operating temperatures. The conductivity of 11ScSZ at 800 °C is almost the same with that of the 8YSZ and MnSZ eletrolytes at 1000 °C. However, the conductivity of 11ScSZ sharply decreases to 0.0001 S/cm. Compared with ScSZ, the conductivity of $Zr_{0.75}Mn_{0.25}O_{2-\delta}$ is obviously higher at intermediate-temperatures. It is worth mentioning that the conductivity value of 0.0144 S/cm is twice higher than 8YSZ and two orders of magnitude higher than that of the ScSZ, and can fulfill the conductivity requirement (~1×10^{-2} S/cm) for the electrolyte at 600 °C. This high conductivity of Mn-stabilized ZrO_2 ceramics has, to the best of our knowledge, not been reported previously.

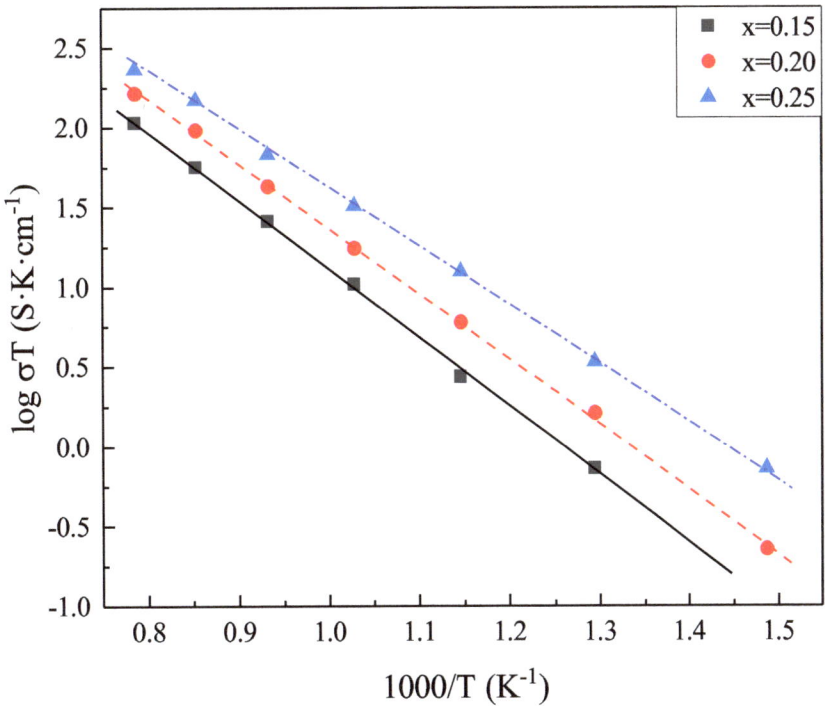

Figure 4. Arrhenius plots for total conductivities with four-probe DC methods.

The activation energy (E_a) values of $Zr_{1-x}Mn_xO_{2-\delta}$ specimens were also listed in Table 1, determined by fitting the curves with the equation:

$$\sigma \cdot T = \sigma_0 \cdot \exp\left(-\frac{E_a}{k \cdot T}\right) \quad (1)$$

Table 1. Specific conductance at different temperatures and activation energy values for $Zr_{1-x}Mn_xO_{2-\delta}$, 8YSZ and 11ScSZ.

Composition	$\sigma_{600°C}$ (S/cm)	$\sigma_{800°C}$ (S/cm)	$\sigma_{1000°C}$ (S/cm)	E_a (eV)
$Zr_{0.85}Mn_{0.15}O_{2-\delta}$	0.0031	0.024	0.085	0.86
$Zr_{0.80}Mn_{0.20}O_{2-\delta}$	0.0069	0.040	0.130	0.81
$Zr_{0.75}Mn_{0.25}O_{2-\delta}$	0.0144	0.064	0.182	0.72
8YSZ [6]	0.0039	0.040	0.140	0.84
11ScSZ [6]	~0.0001	0.112	0.302	0.70

The Arrhenius plot with linear behavior indicates that $Zr_{1-x}Mn_xO_{2-\delta}$ compounds are highly ion conducting and their conductivity increases with increasing temperature and MnO doping content. The activation energy value of the $Zr_{1-x}Mn_xO_{2-\delta}$ ceramics decreases with the increasing Mn doping content. The higher the concentration of Mn, the higher the migration rate of the oxygen vacancy charge carrier. The lowest activation energy (0.72 eV) is lower than that for YSZ (0.84 eV), but higher than that for ScSZ (0.70 eV). It means that the ionic migration in Mn-stabilized ZrO_2 is much easier than in Y-stabilized ZrO_2, but harder than in Sc-stabilized ZrO_2.

4. Conclusions

Cubic phase Mn-stabilized ZrO_2 samples containing 15, 20, and 25 at.% Mn were synthesized using a traditional solid state reaction method at 1400 °C for 12 h. A detailed microstructural characterization employing the Rietveld refinement method reveals that Mn ions entered the ZrO_2 crystal lattice in the form of Mn^{2+} and $Zr_{1-x}Mn_xO_{2-\delta}$ are composed of single cubic ZrO_2 phase. Both the grain size and oxygen vacancy concentration increased with th increasing Mn doping content. The decrease in the lattice parameter corroborates with the increase in oxygen vacancy with the increase in MnO concentration in the ZrO_2 lattice. Ionic conductivity of all $Zr_{1-x}Mn_xO_{2-\delta}$ are found to increase with the increasing Mn content as well as temperature. The highest conductivity was observed from the $Zr_{0.75}Mn_{0.25}O_{2-\delta}$ samples, which reached 0.0144 S/cm and 0.182 S/cm at 600 °C and 1000 °C, respectively. The obtained results proved that this compound is promising for IT-SOFC application. The future scope of this work includes researching the ionic and electron conductivity of the Mn-stabilized ZrO_2 system more systematically, identifying a method for grain refinement and evaluating the stability of the resulting microstructure and performance.

Author Contributions: Validation, S.Z. and L.J.; formal analysis, H.Z.; investigation, R.G.; data curation, C.J. and Y.W.; writing—review and editing, L.G.; supervision, J.W. All authors have read and agreed to the published version of the manuscript.

Funding: This research received no external funding.

Institutional Review Board Statement: Not applicable.

Informed Consent Statement: Not applicable.

Data Availability Statement: Not applicable.

Acknowledgments: This work was financially supported by the National Natural Science Foundation of China (Grant No. 51502235 and Grant No. 51777172), Natural Science Foundation of Shaanxi Province (Grant No. 2021JQ-884) and Science and Technology Research Project of Yingkou Institute of Technology (Grant No. 110505010).

Conflicts of Interest: The authors declare no conflict of interest.

References

1. Zakaria, Z.; Kamarudin, S.K. Enhancement on the Quaternized sodium alginate/polyvinyl alcohol membrane performance in the application of passive DEFCs. *Mater. Lett.* **2022**, *309*, 131388. [CrossRef]
2. Bonnet, E.; Grenier, J.C.; Bassat, J.M.; Jacob, A.; Delatouche, B.; Bourdais, S. On the ionic conductivity of some zirconia-derived high-entropy oxides. *J. Eur. Ceram. Soc.* **2021**, *41*, 4505–4515. [CrossRef]
3. Abdalaa, P.M.; Lamas, D.G.; Fantini, M.C.A.; Craievich, A.F. Retention at room temperature of the tetragonal t"-form in Sc_2O_3-doped ZrO_2 nanopowders. *J. Alloys Compd.* **2010**, *495*, 561–564. [CrossRef]
4. Sarat, S.; Sammes, N.; Smirnova, A. Bismuth oxide doped scandia-stabilized zirconia electrolyte for the intermediate temperature solid oxide fuel cells. *J. Power Sources* **2006**, *160*, 892–896. [CrossRef]
5. Shukla, V.; Kumar, A.; Basheer, I.L.; Balani, K.; Subramaniam, A.; Omar, S. Structural Characteristics and Electrical Conductivity of Spark Plasma Sintered Ytterbia Co-doped Scandia Stabilized Zirconia. *J. Am. Ceram. Soc.* **2017**, *100*, 204–214. [CrossRef]
6. Arachi, Y.; Sakai, H.; Yamamoto, O.; Takeda, Y.; Imanishai, N. Electrical conductivity of the $ZrO_2 - Ln_2O_3$ (lanthanides) system. *Solid State Ionics* **1999**, *121*, 133–139. [CrossRef]
7. Dravid, V.P.; Ravikumar, V.; Notis, M.R.; Lyman, C.E.; Dhalenne, G.; Revcolevschi, A. Stabilization of Cubic Zirconia with Manganese Oxide. *J. Am. Ceram. Soc.* **1994**, *77*, 2758–2762. [CrossRef]
8. Valigi, M.; Gazzoli, D.; Dragone, R.; Marucci, A.; Matteib, G. Manganese oxide-zirconium oxide solid solutions. An X-ray diffraction, Raman spectroscopy, thermogravimetry and magnetic study. *J. Mater. Chem.* **1996**, *6*, 403–408. [CrossRef]
9. Choudhary, V.R.; Uphade, B.S.; Pataskar, S.G.; Keshavaraja, A. Low-temperature complete combustion of methane over Mn-, Co-, and Fe-stabilized ZrO_2. *Angew. Chem. Int. Ed. Engl.* **1996**, *35*, 2393–2395. [CrossRef]
10. Ostanin, S.; Ernst, A.; Sandratskii, L.M.; Bruno, P.; Däne, M.; Hughes, I.D.; Staunton, J.B.; Hergert, W.; Mertig, I.; Kudrnovský, J. Mn-stabilized zirconia: From imitation diamonds to a new potential high-T_C ferromagnetic spintronics material. *Phys. Rev. Lett.* **2007**, *98*, 016101. [CrossRef]
11. Pucci, A.; Clavel, G.; Willinger, M.G.; Zitoun, D.; Pinna, N. Transition Metal-Doped ZrO_2 and HfO_2 Nanocrystals. *J. Phys. Chem. C* **2009**, *113*, 12048–12058. [CrossRef]

12. Zakaria, Z.; Kamarudin, S.K.; Timmiati, S.N. Influence of Graphene Oxide on the Ethanol Permeability and Ionic Conductivity of QPVA-Based Membrane in Passive Alkaline Direct Ethanol Fuel Cells. *Nanoscale Res. Lett.* **2019**, *14*, 28. [CrossRef] [PubMed]
13. Pal, S.; Mondal, R.; Guha, S.; Chatterjee, U.; Jewrajka, S.K. Homogeneous phase crosslinked poly(acrylonitrile-co-2-acrylamido-2-methyl-1-propanesulfonic acid) conetwork cation exchange membranes showing high electrochemical properties and electrodialysis performance. *Polymer* **2019**, *180*, 121680. [CrossRef]
14. Kawada, T.; Sakai, N.; Yokokawa, H.; Doklya, M. Electrical properties of transition-metal-doped YSZ. *Solid State Ionics* **1992**, *53*, 418–425. [CrossRef]
15. Pomykalska, D.; Bućko, M.M.; Rekas, M. Electrical conductivity of MnO_x-Y_2O_3-ZrO_2 solid solutions. *Solid State Ionics* **2010**, *181*, 48–52. [CrossRef]
16. Kim, J.H.; Choi, G.M. Mixed ionic and electronic conductivity of $[(ZrO_2)_{0.92}(Y_2O_3)_{0.08}]_{1-y} \cdot (MnO_{1.5})_y$. *Solid State Ionics* **2000**, *130*, 157–168. [CrossRef]
17. Lei, Z.; Zhu, Q.S. Phase transformation and low temperature sintering of manganese oxide and scandia co-doped zirconia. *Mater. Lett.* **2007**, *61*, 1311–1314. [CrossRef]
18. Nandy, A.; Dutta, A.; Pradhan, S.K. Effect of Manganese (II) Oxide on microstructure and ionic transport properties of nanostructured cubic zirconia. *Electrochem. Acta* **2015**, *170*, 360–368. [CrossRef]
19. Gao, L.; Xie, M.J.; Jin, L.H.; Wang, Y.; Jin, C.Q.; Zhao, Y.H. Mn-stabilized zirconia ceramics: Phase transformation and mixed ionic-electronic conductivity. *Ceram. Int.* **2018**, *44*, 19383–19389. [CrossRef]
20. Gao, L.; Zhou, L.; Li, C.S.; Feng, J.Q.; Lu, Y.F. Kinetics of stabilized cubic zirconia formation from MnO_2-ZrO_2 diffusion couple. *J. Mater. Sci.* **2013**, *48*, 974–977. [CrossRef]
21. Herle, J.V.; Vasquez, R. Conductivity of Mn and Ni-doped stabilized zirconia electrolyte. *J. Eur. Ceram. Soc.* **2004**, *24*, 1177–1180. [CrossRef]
22. Rahaman, M.A.; Rout, S.; Thomas, J.P.; McGillivary, D.; Leung, K.T. Defect-rich dopant-free ZrO_2 nanostructures with superior dilute ferromagnetic semiconductor properties. *J. Am. Chem. Soc.* **2016**, *138*, 11896–11906. [CrossRef] [PubMed]
23. Renuka, L.; Anantharaju, K.S.; Sharma, S.C.; Nagaswarupa, H.P.; Prashantha, S.C.; Nagabhushana, H.; Vidya, Y.S. Hollow microspheres Mg-doped ZrO_2 nanoparticles: Green assisted synthesis and applications in photocatalysis and photoluminescence. *J. Alloys Compd.* **2016**, *672*, 609–622. [CrossRef]

Article

Study of Crack-Propagation Mechanism of $Al_{0.1}CoCrFeNi$ High-Entropy Alloy by Molecular Dynamics Method

Cuixia Liu [1,*] and Yu Yao [2]

[1] School of Materials Science and Chemical Engineering, Xi'an Technological University, Xi'an 710021, China
[2] Jiangsu Longda Superalloy Co., Ltd., Wuxi 214105, China
* Correspondence: cuixialiu2016@sina.com; Tel.: +86-29-152-9145-7058

Abstract: The crack propagation mechanism of $Al_{0.1}CoCrFeNi$ high-entropy alloy (HEA) was investigated with the molecular dynamics method. The pre-crack propagation and stretching processes of single-crystal $Al_{0.1}CoCrFeNi$ HEA and $Al_{0.1}CoCrFeNi$ HEA with grain boundaries were simulated. The effects of strain rates and different crystal structures on the crack propagation of the alloy therein at room temperature were studied. They both exhibited plastic deformation and ductile fracturing, and the crack tip involved dislocations at 45° and 135° under the tensile stress. The dislocations formed in the intrinsic-stacking fault and stacking fault based on hexagonal closely packed structures spread and then accumulated near the grain boundary. At the position where hexagonal closely packed structures were accumulated, the dent was obviously serious at the 1/3 position of the alloy where the fracturing finally occurred. The yield strength for $Al_{0.1}CoCrFeNi$ HEA with grain boundaries was lower than that of the single-crystal $Al_{0.1}CoCrFeNi$ HEA. However, Young's moduli for $Al_{0.1}CoCrFeNi$ HEA with grain boundaries were higher than those of the single-crystal $Al_{0.1}CoCrFeNi$ HEA. The grain boundaries can be used as the emission source of dislocations, and it is easier to form dislocations in the-single crystal $Al_{0.1}CoCrFeNi$ HEA, but the existence of grain boundaries hinders the slippage of dislocations.

Keywords: high-entropy alloy; crack propagation; molecular dynamics; dislocation

Citation: Liu, C.; Yao, Y. Study of Crack-Propagation Mechanism of $Al_{0.1}CoCrFeNi$ High-Entropy Alloy by Molecular Dynamics Method. *Crystals* **2023**, *13*, 11. https://doi.org/10.3390/cryst13010011

Academic Editors: Linghang Wang and Gang Xu

Received: 21 November 2022
Revised: 9 December 2022
Accepted: 14 December 2022
Published: 22 December 2022

Copyright: © 2022 by the authors. Licensee MDPI, Basel, Switzerland. This article is an open access article distributed under the terms and conditions of the Creative Commons Attribution (CC BY) license (https://creativecommons.org/licenses/by/4.0/).

1. Introduction

A high-entropy alloy (HEA) is multi-component alloy with equiatomic or near-equiatomic composition, which describes a solid solution composed of five or more metal elements. The molar ratios of these elements are similar or equal, usually ranging from 5% to 35% [1]. The excellent mechanical properties of $Al_{0.1}CuFeNi$ HEA have attracted a large number of scholars to study it deeply because of its remarkable mechanical strength, oxidation resistance, and fatigue resistance at high temperatures [2,3]. In its plastic and elastic process, crack expansion is a typical mechanical behavior, one that is characteristic of HEAs, and it has an important influence on its mechanical properties. Cheng [4] simulated and calculated fracture toughness by molecular dynamics, which indicated that fracture toughness measured in stress intensity factor (or energy release rate) decreases with the decreasing crack length. The difficulties in preparing $Al_{0.1}CrCuFeNi$ HEA prevented people from moving forward, and the mechanism of its plastic deformation is still unclear. Therefore, it is a significant breakthrough to study the microscopic evidence of the mechanical properties of $Al_{0.1}CrCuFeNi$ HEA during crack expansion. During the plastic deformation of an HEA, near the crack tip, each dislocation nucleates and continues via emission. At the same time, stacking layers, or twins, are formed. Therefore, atomic shear behavior is caused by dislocation motion.

However, it is challenging to research the crack propagation mechanism for $Al_{0.1}CrCuFeNi$ HEA. There are many factors that affect crack propagation, such as crystal structure, dislocation slip direction, elemental composition, and manufacturing processes. In early work on the topic, our research team [5–7] discussed the effect of grain boundaries and short-range order on the mechanical properties of HEAs and explained the deformation mechanism. Zhang Z J et al. researched the local cracks on the surface and inside of an HEA with a FCC crystal structure [8]. The two cracks were roughly parallel to and extended along the same crystal direction; they were connected by a trace. The angle between the crack direction and the trace was about 71°, which is exactly the angle between the two (111) slip surfaces of HEA with FCC. The above results show that there is a certain relationship between crack propagation and dislocation slip in HEA with FCC.

Molecular dynamics is an effective method for exploring the essential mechanism of the mechanical properties of HEAs [9]. Li investigated $Co_{25}Ni_{25}Fe_{25}Al_{7.5}Cu_{17.5}$ HEAs by a series of molecular-dynamics tensile tests at different strain rates and temperatures [10]. Wang studied the mechanical behaviors and deformation mechanisms of scratched AlCrCuFeNi HEAs by molecular-dynamics simulations [11]. The movement trend of dislocation and stacking faults in crack expansion may be analyzed at the microscopic scale. In this study, molecular-dynamics was used to simulate the influence of loading rate on the propagation process of pre-cracks in an $Al_{0.1}CoCrFeNi$ HEA. The rest of paper is organized as follows: The description of the simulation method is offered in Section 2. Section 3 provides the discussion of the results of the simulation. The concluding remarks via the simulation results are given in Section 4.

2. Computational Methods

The interactions between atoms of matter is measured by a potential function which includes the embedded atom method (EAM) [12], Lennard–Jones potential [13], Mores potential [14], Johnson potential [15], and so on. Among them, the EAM potential has been applied to study the kinetics and solidification processes of liquid metal, surface structures, adsorption and microcosmic clusters, and many other things with great success. It can be expressed by the following:

$$E_i = F_\alpha \left(\sum_{j \neq i} \rho_\beta(r_{ij}) \right) + \frac{1}{2} \sum_{j \neq i} \phi_{\alpha\beta}(r_{ij}) \quad (1)$$

where, E_i is the sum of embedded energies and counter-potentials of two systems; α and β represent the element types of atom i and j; r_{ij} is the distance between atom i and j; F represents the embedding energy function, which is the function of the electron density of all atoms in the system except itself ρ; $\Phi_{\alpha\beta}$ represents the potential interaction between elements α and β, which is a function of the distance r_{ij} between atoms i and j.

Based on the EAM theory, Professor Xia investigated the potential function of the AlCoCrFeNi system for several years and obtained an accurate potential function [16]. In this study, this potential function was used and also verified by us.

The models for $Al_{0.1}CoCrFeNi$ HEAs with single-crystal (SC-HEA) and $Al_{0.1}CoCrFeNi$ HEAs with grain boundaries (GB-HEA) are shown in Figure 1. Figure 1a shows a sketch map of a pre-crack for the SC-HEA. The standard of the International Organization for Standardization is not valid at the nanoscale, and therefore, the angle of the pre-crack may be random [17]. In this study, 45° was chosen. The size and orientation of the model in three-dimensional space (X, Y and Z) are given by 100a0 × 35a0 × 35a0; and $[11\bar{2}]$, [111], and $[1\bar{1}0]$, respectively. The lattice parameter a0 was 3.581Å. Finally, the SC-HEA was obtained as shown in Figure 1b,c. Compared with the SC-HEA, GB-HEA including grain boundaries (GB) contained a clustered microstructure, which was established based on the SC-HEA. The SC-HEA was uniformly heated from 300 to 2400 K and then relaxed at 2400 K in order to make all atoms keep a liquid state. After that, the $Al_{0.1}CoCrFeNi$ HEA was cooled down to room temperature at a certain cooling rate to form the cluster microstructure shown in Figure 1d,e. It can be seen that after solidification, $Al_{0.1}CoCrFeNi$ HEA still kept a FCC crystal structure. A small number of HCP and BCC atoms were precipitated after the polycrystalline model's relaxation by the crystal boundary.

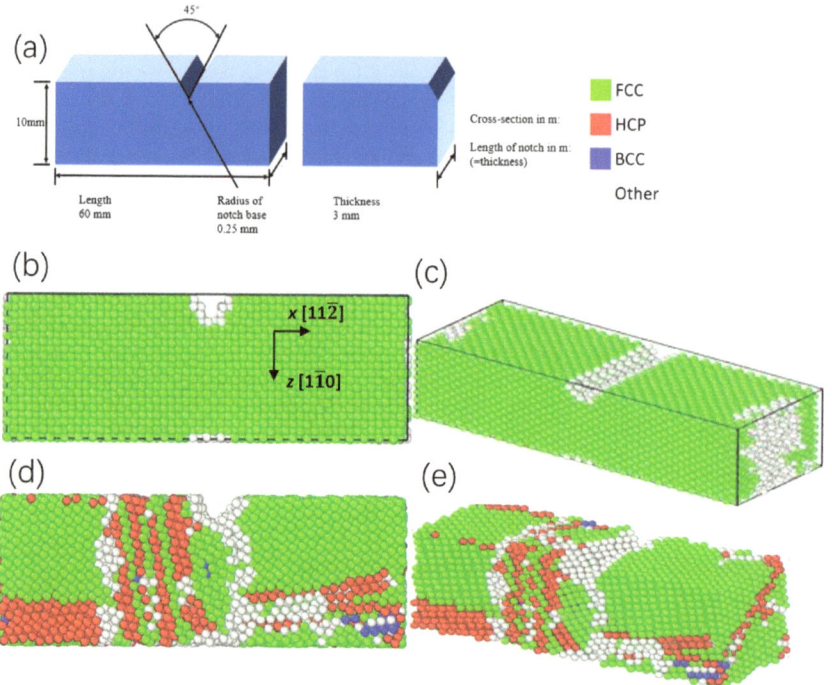

Figure 1. Model for a pre-crack in Al$_{0.1}$CoCrFeNi HEA—single crystal and grain boundary versions: (**a**) sketch map of pre-crack for SC-HEA; (**b**,**c**) the SC-HEA model; (**d**,**e**) the GB-HEA model.

3. Results and Discussion

Based on above models of SC-HEA and GB-HEA, the crack's direction is $(11\bar{2})[1\bar{1}0]$. It is important to choose an appropriate strain rate when Al$_{0.1}$CrCuFeNi HEA is stretched in the molecular dynamics method. If the strain rate is too low, it would require a very long simulation time. If the strain rate is too high, it would result in higher tensile strength. Therefore, the strain rate should not only make valence bonds between atoms fractured, but also take into account the time allowed for the computer simulation. If the strain rate is less than 10^8 s^{-1}, it takes too long to simulate. If it is higher than 10^{10} s^{-1}, the tensile strength is too high. In this paper, SC-HEA and GB-HEA were stretched at a constant quasi-static loading speed along the x axis with different strain rates, including 5×10^8, 8×10^8 s^{-1}, 1×10^9, and 2×10^9 s^{-1} and 5×10^9 s^{-1} at 300 K in order to investigate the microstructural evolution of crack propagation.

3.1. Crack Propagation for SC-HEA

The evolution of crack propagation microstructurally in SC-HEA is shown in Figure 2. Overall, the deformation processes of crack propagation in SC-HEA with different strain rates are similar. Take the deformation process of 8×10^8 s^{-1} (Figure 2b) as an example: Before $\varepsilon = 0.034$, it is in the initial stage of crack propagation; the applied load has not yet acted on the crack-tip area, and the concentrated stress on the crack tip has not yet reached the critical value, resulting in the crack width increasing along the x axis and the length remaining unchanged. It can be seen in Figure 2b that when $\varepsilon = 0.034$, the crack tip begins to propagate, and the critical stress at the crack tip is 5.45 GPa (shown in Figure 3b). The arrow in the figure shows the first emitted dislocation. At this time, FCC is the main feature of the atomic crystal structure of Al$_{0.1}$CoCrFeNi HEA. When $\varepsilon > 0.034$, the crack tip begins to distort gradually, and the crack starts to propagate rapidly along the $[1\bar{1}0]$ direction. When $\varepsilon = 0.053$, the atoms near the crack tip (area A) move in disorder. As the crack tip emits dislocations many times, passivation occurs to a certain extent, resulting in the stress always being concentrated on the crack tip. The dislocations emitted from the crack tip move along the direction of the slip plane in Figure 2b, and a dislocation-free area (area B) is formed between the movement track and the crack tip. The stress is redistributed at the front end of the crack tip after the dislocation is emitted. The new deformation mechanisms appear in the crack-tip-area dislocation and dislocation ring, which greatly release the stress concentration phenomenon into the system. When $\varepsilon = 0.085$, due to the continuous emission of dislocations, there are layered, hexagonal, closely packed (HCP) atoms (red atoms) and a small number of body-centered cubic (BCC) atoms (blue atoms). The HCP structure may form a stacking fault (SF). When $\varepsilon = 0.304$, the necking phenomenon appears gradually, which indicates that the deformation process of SC-HEA belongs to plastic deformation. After that, the SC-HEA enters an uneven plastic deformation stage, during which twins and stacking faults keep growing and disappearing. The system is about to break when ε is 0.521. It is judged that SC-HEA exhibits ductile fracturing [18]. When the strain rate increases from 5×10^8 s^{-1} to 5×10^9 s^{-1}, when the fracture occurs, the strain value (ε) fluctuates but mostly reduces, giving values such as 0.630, 0.521, 0.540, 0.512, and 0.533. In the process of crack propagation, the stress always concentrates on the crack tip, which leads to obvious passivation of the crack tip. Dislocations are intermittently emitted from the crack tip along two symmetrically distributed slip surfaces, and the inclination angles of the two symmetrically distributed slip surfaces are 45° and 135°, shown in Figure 2. The dislocation moves forward along the slip plane to pile up at the location of boundary, which causes a great stress concentration where dislocations accumulate. As the stretching continues to increase, the dislocation will cross the boundary and extend into neighboring grains. Therefore, plastic deformation would occur until the SC-HEA fractures. At the same time, vacancies would be formed and the extrinsic stacking fault (ESF) will appears, as shown in Figure 2b,c,e. The structure for ESF is always that two HCP layers are sandwiched between FCC layers during deformation, which is consistent with the theoretical analysis [19] and the experimental results [20]. During the plastic deformation of AlCoCrFeNi$_{2.1}$ HEA in an in situ tensile test monitored by a transmission electron microscope, dislocations were emitted from FCC phase and piled up in large quantities at the right phase boundary.

The stress–strain curves of SC-HEA under different strain rates are shown in Figure 3. It can be seen that there is an initial linear elastic stage at first, and a linear equation was fitted; see Figure 3. The stress reached its peak, and then the stress dropped sharply until fracture. After that, the deformation of SC-HEA entered the plastic deformation stage. With increasing strain, SC-HEA continued deform until it fractured. When the strain rate increased from 5×10^8 to 5×10^9 s^{-1}, the strain ε corresponding to the peak value of yield stress was 0.0423, 0.0550, 0.0406, 0.0371, or 0.0432, respectively, and the speed of the peak value in the tensile process of the system fluctuated.

Table 1 lists the changes in yield strength and Young's modulus with strain rate. It can be observed that with an increase in strain rate, the yield strength and Young's modulus both increase, although they had little fluctuation. This indicates that a higher strain rate for SC-HEA will cause higher yield strength and a higher Young's modulus.

Table 1. Young's modulus and yield strength of SC-HEA at different strain rates.

Strain Rate/s^{-1}	5×10^8	8×10^8	1×10^9	2×10^9	5×10^9
Yield strength/GPa	5.7786	5.9517	5.7458	6.1128	6.4878
Young's modulus/GPa	66.8081	52.4974	69.2321	89.7900	74.7483

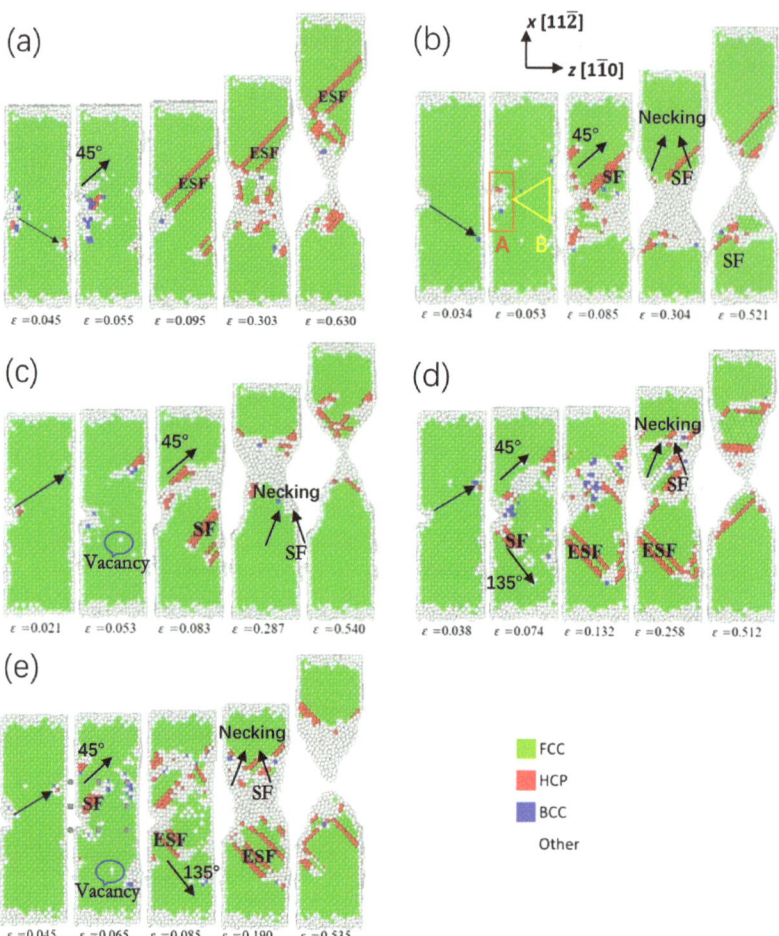

Figure 2. Crack propagation process of SC-HEA in tensile tests at different strain rates: (**a**) 5×10^8 s^{-1} (**b**) 8×10^8 s^{-1} (**c**) 1×10^9 s^{-1} (**d**) 2×10^9 s^{-1} (**e**) 5×10^9 s^{-1}.

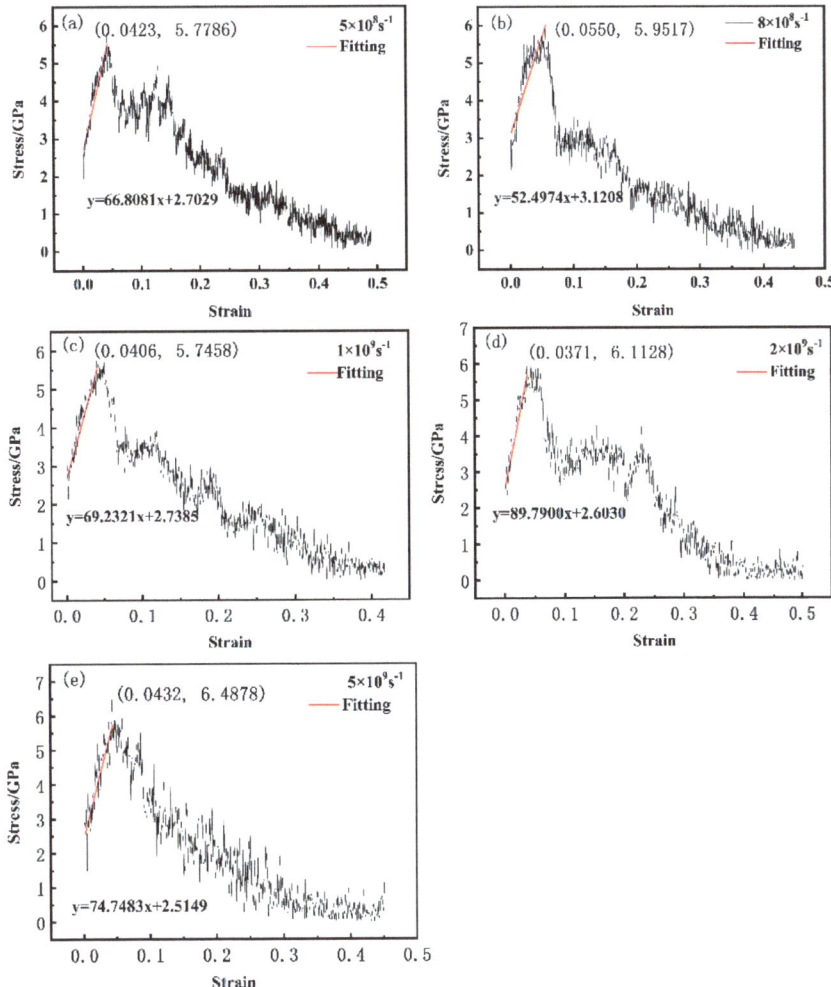

Figure 3. Stress–strain curves of SC-HEA at different strain rates at 300 K: (**a**) 5×10^8 s^{-1} (**b**) 8×10^8 s^{-1} (**c**) 1×10^9 s^{-1} (**d**) 2×10^9 s^{-1} (**e**) 5×10^9 s^{-1}.

3.2. Crack Propagation for GB-HEA

Compared with the crack propagation process of SC-HEA, that of GB-HEA is shown in Figure 4. The deformation process of GB-HEA is similar to that of SC-HEA, which is also plastic deformation. However, GB-HEA contains GB, and the motion of the dislocation is clearly affected by GB. Take the deformation process of 8×10^8 s^{-1} (Figure 4b) as an example. In Figure 4b, when $\varepsilon = 0.034$, because there are GB under the crack tip, the strain field expands from the crack tip and interacts with the GB. No dislocation occurs at the crack tip, which proves GB may impede the propagation of cracks. When $\varepsilon = 0.053$, dislocations are emitted from the GB (area A). When $\varepsilon = 0.085$, deformations at area A continue to extend to the interior of HEA slowly. At the crack tip, the GB structure still hinders dislocation, which results in more dislocations at the GB. SF are formed in the middle area. When $\varepsilon = 0.304$, the strain field at the crack tip continues to act with the GB, and the stress concentration phenomenon occurs. The number of disordered atoms (white atoms) increases with the amount of deformation. At this time, the slip band will find it difficult to alleviate this stress concentration phenomenon, resulting in serious damage to the internal structure. This shows that the depression of GB-HEA is obviously serious at the place where the HCP structure accumulates in the up 1/3 position. The intrinsic stacking fault (ISF) is formed during the tensile process, which is shown as two adjacent HCP crystal planes when $\varepsilon = 0.304$. When $\varepsilon = 0.521$, the system shows an obvious necking phenomenon. The crack tip gradually merges with the GB. The twin boundary (TB) appears as a single-layer HCP crystal structure. In general, TB is achieved by moving adjacent atomic planes based on the ISF structure. After the ISF structure is blocked during the extension, in order to ensure the continuous improvement of material plasticity, the ISF develops into a TB. On the one hand, the TB adjusts the crystal orientation, and then further stimulates the structure slip. On the other hand, the twin boundary reduces the average free path of the dislocation and enhances the strain hardening and improves the plasticity, which is consistent in with the TWIP effect in reference [21].

According to Figure 4, when the strain rate increases from 5×10^8 to 5×10^9 s^{-1}, when the fracture occurs, the strain (ε) fluctuates—0.575, 0.617, 0.705, 0.662, and 0.630, respectively, most of these values being higher than those of SC-HEA.

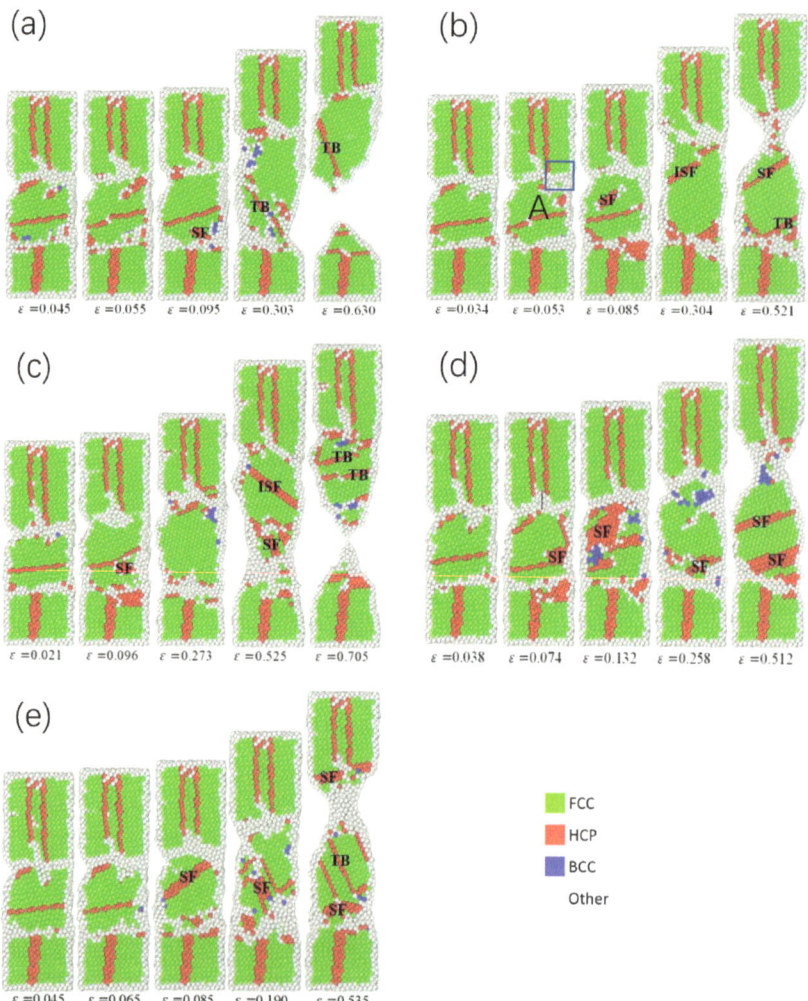

Figure 4. Crack propagation process of cracked GB-HEA in tensile tests with different strain rates: (a) 5×10^8 s^{-1} (b) 8×10^8 s^{-1} (c) 1×10^9 s^{-1} (d) 2×10^9 s^{-1} (e) 5×10^9 s^{-1}.

The stress–strain curves of GB-HEA under different strain rates are shown in Figure 5. In relation to the stress–strain curves of SC-HEA, they have the same variation trends. They both go through a linear elastic region to the peak stress, and the stress drops rapidly after reaching the peak. After that, they enter the plastic deformation stage. When the strain rate increases from 5×10^8 to 5×10^9 s^{-1}, the strain ε when the yield stress reaches the peak value is 0.0302, 0.0362, 0.0350, 0.0378, or 0.0361, respectively.

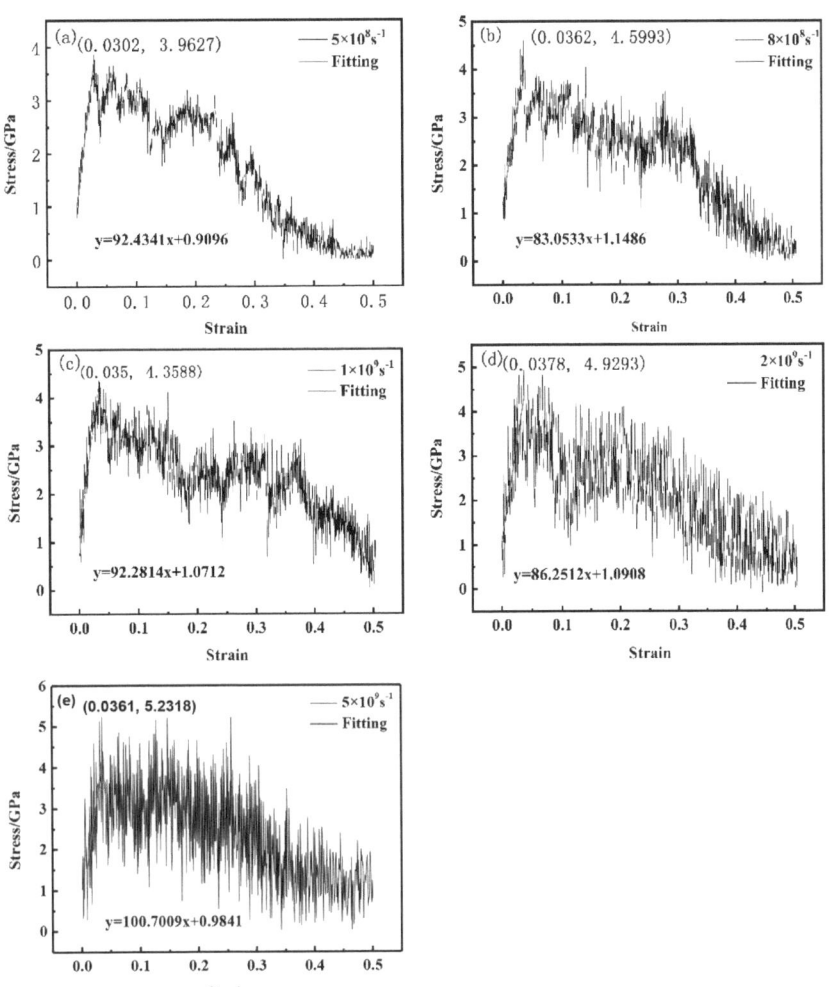

Figure 5. Stress–strain curves of GB-HEA at different loading rates at 300 K: (a) 5×10^8 s^{-1} (b) 8×10^8 s^{-1} (c) 1×10^9 s^{-1} (d) 2×10^9 s^{-1} (e) 5×10^9 s^{-1}.

At the same time, the yield strengths for GB-HEA shown in Table 2 are lower than those of SC-HEA. However, the Young's moduli for GB-HEA are higher than those of SC-HEA, although the Young's modulus for GB-HEA fluctuated a little. Those characteristics all prove that GB may hinder the movement of dislocation. The yield strengths and Young's moduli are compared in Figure 6. Compared with SC-HEA, GB can be used as the emission source of dislocations, which means that by GB, the number of initial slip systems increases, the plastic deformation capacity of the material increases, the stress required for material yield decreases, and the yield strength also decreases. On the other hand, the Young's moduli of the GB-HEA are higher than those of SC-HEA.

Table 2. Young's modulus and yield strength of GB-HEA at different strain rates.

Strain Rate/s^{-1}	5×10^8	8×10^8	1×10^9	2×10^9	5×10^9
Yield strength/GPa	3.9627	4.5993	4.3588	4.9293	5.2318
Young's modulus/GPa	92.4341	83.0533	92.2814	86.2512	100.7009

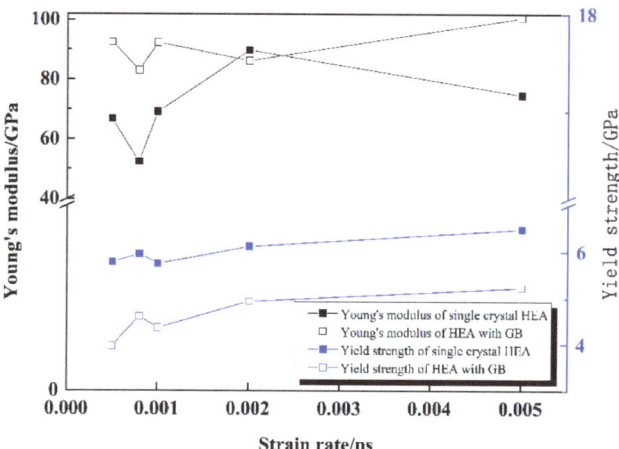

Figure 6. Comparison of Young's modulus and yield strength between single crystal Al$_{0.1}$CoCrFeNi HEA and Al$_{0.1}$CoCrFeNi HEA with grain boundary.

4. Conclusions

SC-HEA and GB-HEA were investigated under different strain rates. The crack propagation mechanisms were analyzed, and the results show that that they both exhibit plastic deformation and ductile fracturing during the tensile process. Dislocations are emitted during the propagation process, the crack tip involves passivation, and the dislocations emitted from the crack tip accumulate near the GB. With the increase in strain rate, the crack propagation rate slows down and the plastic strength of the material increases. The cracks of Al$_{0.1}$CoCrFeNi HEA emit dislocations along the directions of 45° and 135° under the tensile stress. The ISF and SF based on HCP were observed during the deformation process. The fracture finally occurs obviously at the 1/3 position. With the increase in strain rate, the yield strengths and Young's moduli for the SC-HEA and GB-HEA fluctuated, but the overall trends were still increasing. The yield strength for GB-HEA was lower than that of SC-HEA. However, the Young's moduli for GB-HEA were higher than those for SC-HEA. The existence of GB can be used as the emission source of dislocations, and it is easier to form dislocations than SC-HEA, but the existence of GB hinders the slippage of dislocations.

Author Contributions: C.L.: methodology, software, validation, investigation, visualization, writing—original draft, writing—review and editing. Y.Y.: software, writing—review and editing. All authors have read and agreed to the published version of the manuscript.

Funding: This research was funded by the Natural Science Foundation of China (grant number 51971166) and Shaanxi Provincial Science and Technology Plan Project (no. 2021JM-430).

Institutional Review Board Statement: Not applicable.

Informed Consent Statement: Not applicable.

Data Availability Statement: Data is contained within the article. The data presented in this study can be seen in the content above.

Conflicts of Interest: The authors declare no conflict of interest.

References

1. Yeh, J.W.; Chen, S.K.; Lin, S.J.; Gan, J.Y.; Chin, T.S.; Shun, T.T.; Chang, S.Y. Nanostructured high-entropy alloys with multiple principal elements: Novel alloy design concepts and outcomes. *Adv. Eng. Mater.* **2004**, *6*, 299–303. [CrossRef]
2. Zhang, Y.; Zuo, T.T.; Tang, Z.; Gao, M.C.; Dahmen, K.A.; Liaw, P.K.; Lu, Z.P. Microstructures and properties of high-entropy alloys. *Prog. Mater. Sci.* **2014**, *61*, 1–93. [CrossRef]
3. Choudhuri, D.; Komarasamy, M.; Ageh, V. Investigation of plastic deformation modes in Al0.1CoCrFeNi high entropy alloy. *Mater. Chem. Phys.* **2018**, *217*, 308–314. [CrossRef]
4. Cheng, S.-H.; Sun, C.-T. Size-dependent fracture toughness of nanoscale structures: Crack-tip stress approach in molecular dynamics. *J. Nanomech. Micromech.* **2014**, *4*, A414001. [CrossRef]
5. Liu, C.-X.; Wang, R.; Jian, Z.-Y. The Influence of Grain Boundaries on Crystal Structure and Tensile Mechanical Properties of Al0.1CoCrFeNi High-Entropy Alloys Studied by Molecular Dynamics Method. *Crystals* **2022**, *12*, 48. [CrossRef]
6. Sun, Z.; Shi, C.; Liu, C.; Shi, H.; Zhou, J. The effect of short-range order on mechanical properties of high entropy alloy Al0.3CoCrFeNi. *Mater. Des.* **2022**, *223*, 111214. [CrossRef]
7. Yang, Y.-C.; Liu, C.-X.; Lin, C.-Y.; Xia, Z.-H. Core Effect of Local Atomic Configuration and Design principles in AlxCoCrFeNi High-Entropy Alloys. *Scr. Mater.* **2020**, *178*, 181–186. [CrossRef]
8. Zhang, Z.; Mao, M.M.; Wang, J.; Gludovatz, B.; Zhang, Z.; Mao, S.X.; Ritchie, R.O. Nanoscale origins of the damage tolerance of the high-entropy alloy CrMnFeCoNi. *Nat. Commun.* **2015**, *6*, 10143. [CrossRef] [PubMed]
9. Liu, C.-X.; Yang, Y.-C.; Xia, Z.-H. Deformation Mechanism in $Al_{0.1}$CoCrFeNi Σ3(111)[1$\bar{1}$0] High Entropy Alloys-Molecular Dynamics Simulation. *RSC Adv.* **2020**, *10*, 27688–27696.
10. Li, L.; Chen, H.; Fang, Q.; Li, J.; Liu, F.; Liu, Y.; Liaw, P.K. Effect of temperature and strain rate on plastic deformation mechanisms of nanocrystalline high-entropy alloys. *Intermetallics* **2020**, *120*, 106741. [CrossRef]
11. Wang, Z.-N.; Li, J.; Fang, Q.-H.; Liu, B.; Zhang, L.-C. Investigation into nanoscratching mechanical response of AlCrCuFeNi high-entropy alloys using atomic simulation. *Appl. Surf. Sci.* **2017**, *416*, 470–481. [CrossRef]
12. Daw, M.-S.; Baskes, M.-I. Embedded-atom method: Derivation and application to impurities, surfaces, and other defects in metals. *Phys. Rev. B* **1984**, *29*, 6443–6453. [CrossRef]
13. Lim, T.-C. The relationship between Lennard-Jones (12–6) and Morse potential functions. *Z. Fur Nat. A J. Phys. Sci.* **2003**, *58*, 615–617. [CrossRef]
14. Soylu, A.; Bayrak, O.; Boztosun, I. Effect of the velocity-dependent potentials on the energy eigenvalues of the Morse potential. *Open Phys.* **2012**, *10*, 953–959. [CrossRef]
15. Roller, D.; Tran, F.; Blaha, P. Merits and limits of the modified Becke-Johnson exchange potential. *Phys. Rev. B* **2011**, *83*, 173–184.
16. Yang, Y.-C.; Liu, C.-X.; Lin, C.-Y.; Xia, Z.-H. The effect of local atomic configuration in high-entropy alloys on the dislocation behaviors and mechanical properties. *Mater. Sci. Eng. A* **2021**, *815*, 141253. [CrossRef]
17. Wang, J.; Yu, L.-M.; Yuan, H.-A.; Li, H.-J.; Liu, Y.-C. Effect of Crystal Orientation and He Density on Crack Propagation Behavior of bcc-Fe. *Acta Metall. Sin.* **2018**, *54*, 47–54.
18. Payne, M.C.; Teter, M.P.; Allan, D.C.; Arias, T.A.; Joannopoulos, A.J. Iterative minimization techniques for abinitio total-energy calculations: Molecular dynamics and conjugate gradients. *Rev. Mod. Phys.* **1992**, *64*, 1045–1097. [CrossRef]
19. Li, W.; Liaw, P.-K.; Gao, Y. Fracture resistance of high entropy alloys: A review. *Intermetallics* **2018**, *99*, 69–83. [CrossRef]
20. Liu, G. In situ TEM Study of AlCoCrFeNi Dual-Phase High-Entropy Alloy. Master's Thesis, Zhejiang University, Hangzhou, China, 2019; pp. 41–47.
21. Bahramyan, M.; Mousavian, R.T.; Brabazon, D. Study of the plastic deformation mechanism of TRIP–TWIP high entropy alloys at the atomic level. *Int. J. Plast.* **2020**, *127*, 102649. [CrossRef]

Disclaimer/Publisher's Note: The statements, opinions and data contained in all publications are solely those of the individual author(s) and contributor(s) and not of MDPI and/or the editor(s). MDPI and/or the editor(s) disclaim responsibility for any injury to people or property resulting from any ideas, methods, instructions or products referred to in the content.

Article

The Influence of Grain Boundaries on Crystal Structure and Tensile Mechanical Properties of Al$_{0.1}$CoCrFeNi High-Entropy Alloys Studied by Molecular Dynamics Method

Cuixia Liu *, Rui Wang and Zengyun Jian

School of Materials Science and Chemical Engineering, Xi'an Technological University, Xi'an 710021, China; ruiwang2019@sina.com (R.W.); jianzengyun@xatu.edu.cn (Z.J.)
* Correspondence: liucuixia@xatu.edu.cn; Tel.: +86-29-15291457058

Abstract: The mechanical properties of high-entropy alloys are superior to those of traditional alloys. However, the key problem of finding a strengthening mechanism is still challenging. In this work, the molecular dynamics method is used to calculate the tensile properties of face-centered cubic Al$_{0.1}$CoCrFeNi high-entropy alloys containing Σ3 grain boundaries and without grain boundary. The atomic model was established by the melting rapid cooling method, then stretched by the static drawing method. The common neighbor analysis and dislocation extraction algorithm are used to analyze the crystal evolution mechanism of Σ3 grain boundaries to improve the material properties of high-entropy alloys during the tensile test. The results show that compared with the mechanical properties Al$_{0.1}$CoCrFeNi high-entropy alloys without grain boundary, the yield strength and Young's modulus of a high-entropy alloy containing Σ3 grain boundary are obviously larger than that of high-entropy alloys without grain boundary. Dislocation type includes mainly 1/6<112> Shockley partial dislocations, a small account of 1/6<110> Stair-rod, 1/2<110>perfect dislocation, and 1/3<111> Hirth dislocations. The mechanical properties of high-entropy alloys are improved by dislocation entanglement and accumulation near the grain boundary.

Keywords: high-entropy alloy; Σ3 grain boundary; molecular dynamics; strengthening mechanism

Citation: Liu, C.; Wang, R.; Jian, Z. The Influence of Grain Boundaries on Crystal Structure and Tensile Mechanical Properties of Al$_{0.1}$CoCrFeNi High-Entropy Alloys Studied by Molecular Dynamics Method. *Crystals* **2022**, *12*, 48. https://doi.org/10.3390/cryst12010048

Academic Editors: Rui Feng and Pavel Lukáč

Received: 26 November 2021
Accepted: 27 December 2021
Published: 30 December 2021

Publisher's Note: MDPI stays neutral with regard to jurisdictional claims in published maps and institutional affiliations.

Copyright: © 2021 by the authors. Licensee MDPI, Basel, Switzerland. This article is an open access article distributed under the terms and conditions of the Creative Commons Attribution (CC BY) license (https://creativecommons.org/licenses/by/4.0/).

1. Introduction

With the rapid development of industry, traditional materials no longer satisfy higher requirements. Research scholars began to pay more attention to high-entropy alloys (HEAs) with high strength, high hardness [1], high wear resistance [2], corrosion resistance [3,4], and suitable thermal stability [5]. HEAs are mixed entropy alloys composed of five or more kinds of elements according to the proportion of equimolar or near equimolar atoms [6]. It generally forms a stable single-phase solid solution, which may be face-centered cubic crystal structure (fcc), body-centered cubic crystal structure (bcc), hexagonal close-packed structure (hcp), or an alloy of two phases [7,8], sometimes nano-phase or even amorphous phase of a class of alloys. Among many kinds of HEAs, the HEAs with fcc crystal structure have considerable stretch plasticity and multiple strengthening mechanisms, and it is gradually becoming the current hot research area in the field of HEAs [9,10]. Thota et al. [11] observed Σ3, Σ9, and Σ27 grain boundaries when they researched the microstructure of as-recrystallized specimens in CoCrFeMnNi HEAs. Chen et al. [12] explains that in FCC alloys, special boundaries were mainly composed of Σ3ns boundaries for CoCrFeMnNi HEAs. Li et al. [13] confirms stacking faults and dislocations present even in the as-homogenized state of the FCC γ matrix in Fe$_{20}$Mn$_{20}$Ni$_{20}$Co$_{20}$Cr$_{20}$ HEAs. Zhao et al. [14] also simulated the influences of Σ3, Σ5, and Σ11 grain boundary on materials performance for CuNiCoFe HEAs. In this paper, Al$_{0.1}$CoCrFeNi HEAs, as typical HEAs as the research object, have a single-phase fcc crystal structure [15]. The deformation process during the tensile test is comparatively studied. For fcc crystal structure, grain boundary energy for Σ3 grain boundary is lowest [16]; as a result, a Σ3 grain boundary is easy to appear in the fcc crystal structure. Therefore, $Al_{0.1}CoCrFeNi$ HEAs with Σ3 grain boundary are studied as a subject.

Among various strengthening methods to improve the mechanical properties of HEAs, grain boundary strengthening is the main effective mechanism. Yang et al. [17] found that the main strengthening mechanism of $Al_{0.1}CoCrFeNi$ high-entropy alloy is fine-grained strengthening and precipitation strengthening. In order to improve the mechanical properties of HEAs, the deformation mechanism during the stretching was accurately analyzed. Traditional experiments focus on the mechanical properties of HEAs mainly. For example, Yang et al. [18] and others studied the Al content of the mechanical properties and found that x is less than 0.5 when HEAs exhibit a single fcc crystal structure. While x is greater than 0.5 and less than 0.9, HEAs display fcc and bcc two-phase mixtures. As the Al content increases, the hardness increases and ductility decreases. For the $Al_{0.1}CoCrFeNi$ HEAs, it is found in the tensile test at different temperatures that the yield strength and ultimate tensile strength decrease with the increase in temperature. The transmission electron microscope shows the normal face-centered cubic slip {111} (110) during the tensile test. In the tensile test at the temperature of 77 K, it was found that the appearance of nano-twins at low temperature can promote the jagged stress-strain curve [19,20]. It is difficult for traditional experiments to describe the micro-evolution process of how grain boundaries in high-entropy alloys improve mechanical properties from the nanoscale. The molecular dynamics simulation can also reduce the elastic and plastic deformation of raw materials to some extent at the atomic scale. It is greatly significant to explain the mechanical properties of crystals and predict the possible deformation mechanism of crystals. Jia et al. [21] studied the mechanical behavior of AlCrFeCuNi HEAs during the tensile deformation using analytical dynamics. It is found that severe lattice distortion and solid solution can lead to dislocation sliding and pinning, thus affecting the mechanical properties of the alloy.

Molecular dynamics simulation is used to stretch the $Al_{0.1}CoCrFeNi$ containing preset Σ3 grain boundary structure at different strain rates and different temperatures in nanoscale. The effect of grain boundary on the excellent mechanical properties and deformation mechanism $Al_{0.1}CoCrFeNi$ HEAs was studied.

2. Computational Methods

In order to study the effect of grain boundary on tensile deformation, the Σ3 grain boundaries of $Al_{0.1}CoCrFeNi$ HEAs with fcc crystal structure are established with molecular dynamics method based on coincidence site lattice (CSL) [22], as shown in Figure 1. The lower the energy of grain boundary, the stabler the crystal structure. As a result, the energy of Σ3 (the orientation angle is 109.47°) grain boundary is lowest in experimental conditions. Σ3 grain boundary selected contains two grains, including GB1 and GB2 (Abbrev. Σ3 GBs HEAs). The model of Σ3 grain boundary is established by obtaining the interface structure by the grain with upper and lower orientation. Through the orient command in the Lammps software (Large-scale Atomic/Molecular Massively Parallel Simulator) to change the X, Y, and Z axes corresponding to the crystal coordinates, on the basis of ensuring that the three directions meet the right-hand rule X, Y, and Z, the coordinates correspond to the works of [11$\bar{2}$], [111], and [101] crystal directions, respectively. Meanwhile, the HEAs model without grain boundary (abbreviation: non-GBs HEAs) was set, as shown in Figure 2. In contrast, the grain orientation of the non-GBs HEAs is consistent with that of GB1.

In order to ensure the representation of the simulation results and the feasibility of the simulation process, the appropriate size should be considered in the modeling process. In order to avoid the rigid docking form of atoms at the boundary and surface caused by periodic boundary conditions and stacking order problems, the boundary length of the model must be carefully selected when defining grain size so as to generate a reasonable model. The size of the model is calculated according to Formulas (1) and (2). The L, B, W are distributed along the Y axis, the X axis, and the Z axis in Figure 1, respectively. For this purpose, the length, width, and height of the three directions correspond to integer times of the minimum periodic constant of the oriented lattice with a size of 62 × 172 × 50.

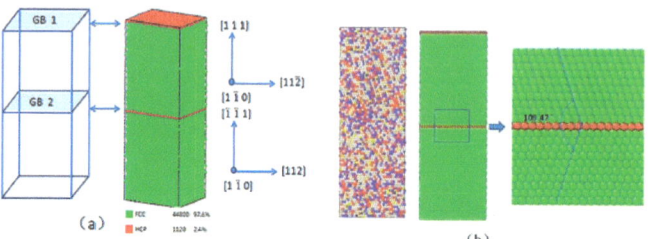

Figure 1. GBs HEAs. (a) 3D structure of GBs HEAs; (b) structure of Σ3{112}.

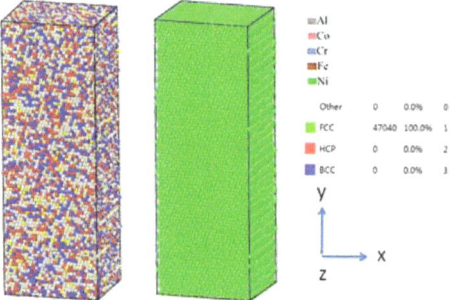

Figure 2. Three-dimensional structure of non-GBs HEAs.

$$L = a \times \sqrt{h^2 + k^2 + l^2} \tag{1}$$

$$B \text{ or } W = b \times \sqrt{h^2 + k^2 + l^2} \tag{2}$$

where h, k, and l are the miller indices along three forward-measured grain directions, respectively. To achieve the desired size, a and b are constants.

To eliminate the boundary effect, periodic boundary conditions are adopted. Because the grain boundary structure itself is unstable structure, static relaxation is needed. After static relaxation, minimizing the energy of the system, the grain boundary structure reaches a stable state. An embedded atom method (EAM) [23] is used to describe the force between atoms, and the timestep is set to 0.001 ps. After the model was built, it was heated above the melting point and cooled at a cooling rate of 1011 K/s to room temperature. As a result, the equilibrium structure was obtained. Figure 3 shows the common neighbor analysis (CNA) of Σ3 GBs HEAs after solidification. We can conclude that the main crystal phase is fcc (97.1%) and a small amount of hcp (2.7%), and bcc (0.1%). The radial distribution function during the solidification process is drawn as shown in Figure 4. The splitting appears in the second peak shape of the curve, which indicates that the system has typical characteristics of a crystalline state. They are consistent with the experimental results [14], which prove the correctness of the model structure and potential function and provide support to predict and analyze the deformation mechanism of $Al_{0.1}CoCrFeNi$ HEAs.

The temperature in the tensile deformation is 100, 300, 600, and 800 K, respectively. The pressure in the X, Z direction remains at standard atmospheric pressure. Different strain rates (1×10^8 s^{-1}, 5×10^8 s^{-1}, 1×10^9 s^{-1}, 5×10^9 s^{-1}) are set for uniaxial tension on the Y axis. The tensile test was performed using a micro-canonical ensemble (NVE) and temp/berendsen ensemble. The molecular dynamics of the model are simulated by Lammps software, and the microstructure evolution of HEAs is analyzed by visual Ovito (Open Visualization Tool) software.

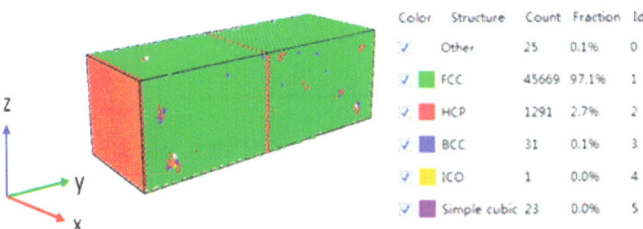

Figure 3. GBs HEAs model after solidification.

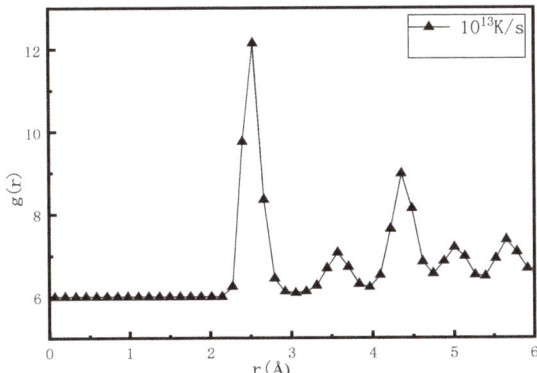

Figure 4. Radial distribution function for Al$_{0.1}$CoCrFeNi HEAs.

The strain rate of the simulated conditions is much larger than that of the actual process. The effects of strain rate and grain boundary on the mechanical properties of HEAs are mainly studied. After each loading, the atomic configuration and stress-strain values in the model are recorded.

During this tensile simulation, the stress-strain value is solved by the Velert equation of motion by velocity Velert method. The corresponding formula of stress-strain is as follows:

$$\sigma = F/A_{xz} \tag{3}$$

$$\varepsilon = (L_y - L_{y0})/L_{y0} \tag{4}$$

where the F is the stress applied along the Y axis. The A_{xz} is the area of the X-Z plane. The L_y is the length of Y direction in the model. L_{y0} is the initial length of Y direction. Each applied strain increment relaxes once at a determined temperature to ensure the grain boundary stability of the HEAs of the fcc crystal structure.

3. Results

The uniaxial stretch was performed on two types of models of HEAs (Σ3 GBs HEAs and non-GBs HEAs) at different strain rates and temperatures to investigate the effect of stacking faults, twins, and dislocations on the mechanical properties of HEAs under the efficiency effect and temperature effect.

3.1. Effect of Σ3 GBs HEAs and Non-GBs HEAs at Different Strain Rates at Room Temperature (300 K)

To study the effect of strain rate on Σ3 GBs HEAs, the uniaxial stretch was selected at a strain rate of 1×10^8 s^{-1}, 5×10^8 s^{-1}, 1×10^9 s^{-1}, and 5×10^9 s^{-1}. The stress-strain curves, microstructure, and dislocation line configurations were simulated and analyzed at 300 K. The effects of Σ3 grain boundaries on stacking faults, dislocation emission, slip, and mechanical properties of high-entropy alloys are studied under different strain rate conditions.

Figure 5 shows the stress-strain curves of Σ3 GBs HEAs and non-GBs HEAs at different strain rates at 300 K. At the elastic stage (that is, the beginning stage), stress-strain curves show a near-linear relationship to the peak yield strength. After that, the system enters a long plastic deformation process until fracture. At the same time, by linear fitting of the linear deformation process of the stress-strain curve with Σ3 GBs HEAs and non-GBs HEAs, Young's modulus is obtained. In classical mechanics, Young's modulus is defined as:

$$e = \sigma/\varepsilon \tag{5}$$

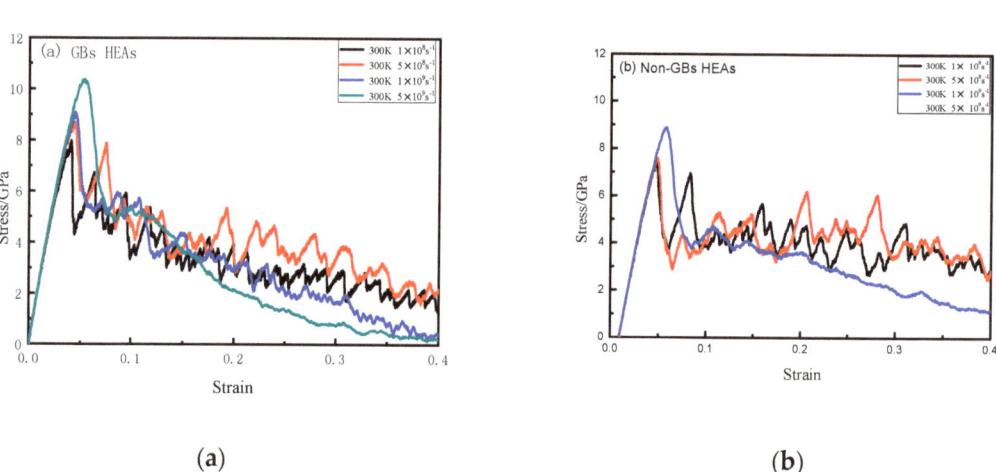

Figure 5. Stress-strain curves of deformation process at different strain rates of HEA. (**a**) GBs HEAs; (**b**) non-GBs HEAs.

The σ is axial stress, and the ε is strain. It is concluded that Young's modulus of Σ3 GBs HEAs is about 230 GPa and that of non-GBs HEAs is about 220 GPa. The density of HEAs is 8151.298 kg/m^3. They were in accord with the experimental values 203 GPa [18] and 7950 kg/m^3 [24]. It also shows that a high strain rate leads to high yield strength. The reason is that during the deformation of the high strain rate, the dislocation motion accelerates, resulting in an increased "near-range resistance", and the deformation of HEAs increases with the increase in strain rate. The experiment [11] also showed that high strain rate sensitivity was observed at high strain rates.

Compare the yield strength and Young's modulus of Σ3 GBs HEAs and non-GBs HEAs. Table 1 shows that under the different strain rates, the yield strength and Young's modulus of Σ3 GBs HEAs are both higher than those of non-GBs HEAs. The corresponding peak stress of non-GBs HEAs is 6.89, 7.51, 7.54, and 8.54 GPa at a strain rate of 1×10^8 s^{-1}, 5×10^8 s^{-1}, 1×10^9 s^{-1}, and 5×10^9 s^{-1} and the strain of 0.040, 0.044, 0.045, and 0.053. However, for Σ3 GBs HEAs, the corresponding peak stress is 7.97, 8.84, 9.03, and 10.34 GPa at a strain rate of 1×10^8 s^{-1}, 5×10^8 s^{-1}, 1×10^9 s^{-1}, and 5×10^9 s^{-1} and the strain of 0.046, 0.049, 0.057, and 0.058, respectively. At the same time, for Σ3 GBs HEAs, the strain corresponding to the peak stress are all higher than that of non-GBs HEAs, which proves that the presence of grain boundary may delay strain and increase the yield strength.

Table 1. Young's modulus and yield strength of GBs HEAs and non-GBs HEAs at different strain rates.

Rate/s^{-1}	Non-GBs Yield Strength/GPa	GBs yield Strength/GPa	Non-GBs Young's Modulus/GPa	GBs Young's Modulus/GPa
1×10^8	6.89	7.97	219.12	230.57
5×10^8	7.51	8.67	219.24	231.63
1×10^9	7.54	9.03	219.71	235.08
5×10^9	8.84	10.34	221.07	235.76

In order to research the morphology of crystal defects in plastic deformation of HEAs, the deformation mechanism of HEAs is analyzed by observing the atomic configuration diagram of high-entropy alloy by CNA diagram. The atomic structure diagrams with strain rates of 1×10^8 s^{-1} and 1×10^9 s^{-1} are selected, respectively, which was shown in Figure 6. Green is the fcc crystal structure, and red is the hcp crystal structure. The single-layer hcp structure is twinned (TB). Two adjacent hcp are intrinsic lamination (ISF), multiple hcp connected together are stacking fault (SF), and one layer between two hcp layers represents extrinsic lamination (ESF) [15].

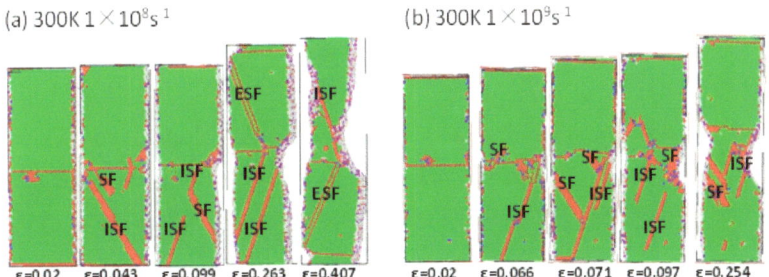

Figure 6. The CNA diagram of GBs HEAs under tensile. (**a**) 300 K, 1×10^8 s^{-1}; (**b**) 300 K, 1×10^9 s^{-1}.

For Σ3 GBs HEAs (Figure 6), the CNA diagram strain rate is 1×108 s-1. When the strain is 0.02, there is an SF near the grain boundary. As the strain increases, ISF emerging observed at strain 0.099, and ISF started moving toward the grain boundary. At a strain of 0.263, ISF moves near to grain boundary, and symmetrical ESF appears on the other side of the grain boundary. The ISF and ESF cross in the opposite direction to form obvious defect regions, which become the crack source of fracture. When the strain rate is 1×10^9 s^{-1}, the CNA diagram under different strains is observed. When the strain is 0.02, there is a stacking fault below the grain boundary. When the strain increases to 0.066, SF moves constantly to the grain boundary and then is blocked by grain boundaries. Therefore, SF is at the grain boundary, and it is difficult to cross grain boundaries. It shows that the grain boundary effectively constrains the movement of the fault or changes the motion direction.

Figure 7a is the CNA diagram of the non-GBs HEAs with a strain rate of 1×10^8 s^{-1}. When the strain is 0.044, SF appears as the strain increases. When the strain is 0.086, an ESF appears throughout the entire model. Figure 7b shows the CNA diagram of the non-GBs HEAs with a strain rate of 1×10^9 s^{-1}. When the strain ranges from 0.071 to 0.075, the ISF and ESF can move through the whole grain. It can be seen from Figure 7 that different types of SF can move in the whole model under different strain rates in non-GBs HEAs. At the same time, the fracture is more difficult to occur in the non-GBs HEAs than in the Σ3 GBs HEAs. In non-GBs HEAs, the movement of the ISF and ESF will not be hindered by grain boundary, so it will not cause the accumulation of stacking fault. In addition, dislocation is also a major factor in influential performance. DXA was used to analyze dislocations with different strain rates, as shown in Figures 8 and 9. Green is 1/6<112> Shockley partial dislocations. Red is 1/6<110> Stair-rod partial dislocations. Blue is 1/2<110>perfect dislocation. Yellow is 1/3<111> Hirth dislocations.

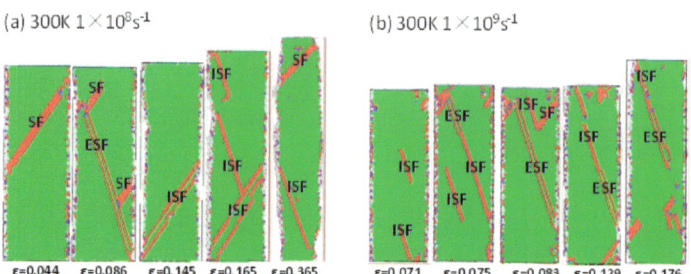

Figure 7. The CNA diagram of non-GBs HEAs under tensile. (a) 300 K, 1×10^8 s^{-1}; (b) 300 K, 1×10^9 s^{-1}.

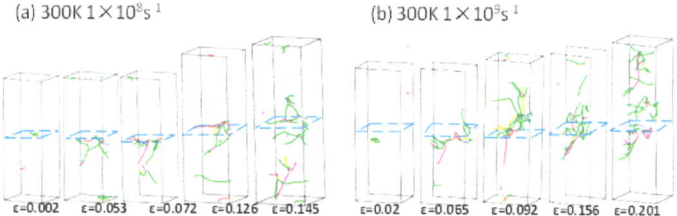

Figure 8. The DXA diagram of GBs HEAs under tensile. (a) 300 K, 1×10^8 s^{-1}; (b) 300 K, 1×10^9 s^{-1}.

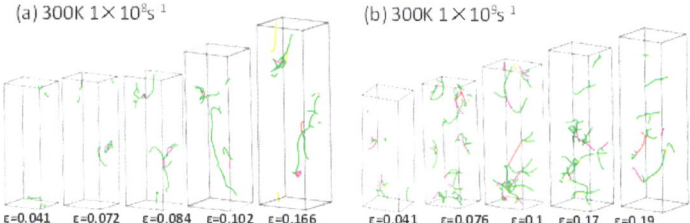

Figure 9. The DXA diagram of non-GBs HEAs under tensile. (a) 300 K, 1×10^8 s^{-1}; (b) 300 K, 1×10^9 s^{-1}.

The main dislocation is of Σ3 GBs HEAs, and non-GBs HEAs is 1/6<112> Shockley partial dislocations in Figures 8 and 9, and the dislocation density of both types of HEAs increases with the increase in strain rate. The strain rate is 1×10^9 s^{-1} (Figure 7). It is found that 1/6<112> Shockley partial dislocations first nucleate at the grain boundary when the strain is 0.02. As the deformation continues, the dislocations proliferate and slip into the crystal, and the crystal begins to deform. At the DXA diagram of strain 0.065, it can be seen that when the dislocation slips to the other side of the grain boundary, the dislocation movement is hindered. Because the atoms on the grain boundary have different orientations, it is difficult to continue to slip and is forced to pile up at the grain boundary. The microstructure of the dislocation line of non-GBs HEAs is shown in Figure 9. The strain rate is 1×10^8 s^{-1}. The strain increases from 0.041 to 0.166, the length of the dislocation increases with the increase in strain. The dislocations in the tensile deformation of the non-GBs can be extended in the whole model without being hindered by grain boundaries.

3.2. Tensile Properties of Σ3 GBs HEAs and Without Non-GBs HEAs at the Same Strain Rate at Different Temperatures

The effect of temperature on the tensile properties of HEAs is very important. Uniaxial tensile experiments are carried out for Σ3 GBs HEAs and non-GBs HEAs with a strain rate of 1×10^8 s^{-1} at 100, 300, 600, and 800 K. By simulating and analyzing the stress and strain curves, micro-atomic and dislocation line configurations during the tensile deformation, the effects of Σ3 GBs HEAs are studied under different temperature conditions.

Figure 10 shows that the stress-strain curves of the deformation process at different temperatures of Σ3 GBs HEAs and non-GBs HEAs. Two curves of stress-strain show similar stress-strain behavior. It can be separated into three stages. The first stage is a linear elastic process. As the deformation continues, the stress rises to its peak and then drops sharply. The third stage is the gradual plastic deformation under low stress until the final structural fracture. At the same time, it is observed that the curves change of the stress-strain curve with Σ3 GBs HEAs after the yield point is more stable than that after the yield point of non-GBs HEAs. Further analysis of internal changes during the tensile deformation is needed.

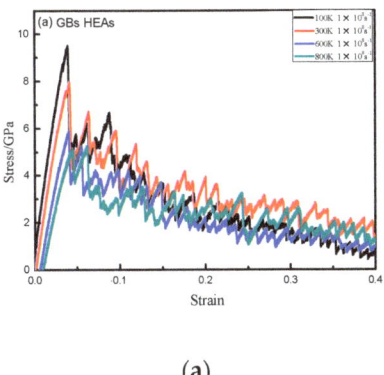
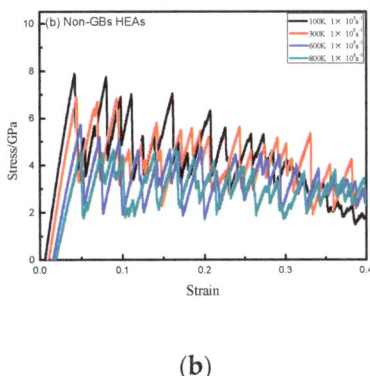

(a) (b)

Figure 10. Stress-strain curves of deformation process at different temperatures of GBs HEAs and non-GBs HEAs; (**a**) GBs HEAs; (**b**) non-GBs HEAs.

Tensile experiments with strain rates of 1×10^8 s^{-1} at temperatures of 100, 300, 600, and 800 K process for Σ3 GBs HEAs and non-GBs HEAs. The yield strength and Young's modulus of both HEAs were obtained from Table 2. As the temperature increased, the yield strength and Young's modulus of Σ3 GBs HEAs were higher than those of non-GBs HEAs. In general, the higher the temperature, the higher the atomic thermal motion, which leads to an unstable structure and decreases the strength. In order to observe obviously the changes of lattice structure at different temperatures, the stacking faults at grain boundaries, and the nucleation and expansion of dislocations in the tensile deformation, the microstructure of atoms and dislocation lines under different strains were selected.

Table 2. Young's modulus and yield strength of GBs HEAs and non-GBs HEAs at different temperatures.

Temperature/K	Non-GBs Yield Strength/GPa	GBsyield Strength/GPa	Non-GBsYoung's Modulus/GPa	GBsYoung's Modulus/GPa
100	7.86	9.47	226.09	242.84
300	6.89	7.97	219.12	230.57
600	5.67	5.86	180.52	188.52

The CNA of Σ3 GBs HEAs and non-GBs HEAs at different temperatures of 100, 300, 600, and 800 K, as observed in Figures 11 and 12. From Figure 11a, when the strain is 0.049, there is an SF above the grain boundary. As the strain increases, the length of ISF above is gradually increasing longer. The grain boundary is also found to hinder the expansion of the SF, ISF, and ESF in the CNA of 300, 600, and 800 K. In the CNA diagram of 100 K non-GBs HEAs (Figure 12). When the strain is 0.056, ISF appears. When the strain is 0.089, ISF runs through the whole model. TB is found at a strain of 0.089. TB appears, which is consistent with the observed phenomena in the experiment [15]. The increasing strain reduced the steady strain hardening of dislocation, which resulted in the greater fluctuation of the stress-strain curve of the non-GBs HEAs. The TB is decomposed from 1/2<110> perfect dislocation to 1/6/<112> Shockley partial dislocations. TB hinders the expansion of the upper right ISF. As the strain increases, found a reduction in the number of ISF, TB has become the main deformation mechanism. As shown in Figure 12b–d, ISF and ESF are not hindered by grain boundary at temperatures of 300, 600, and 800 K and can be expanded in the whole model.

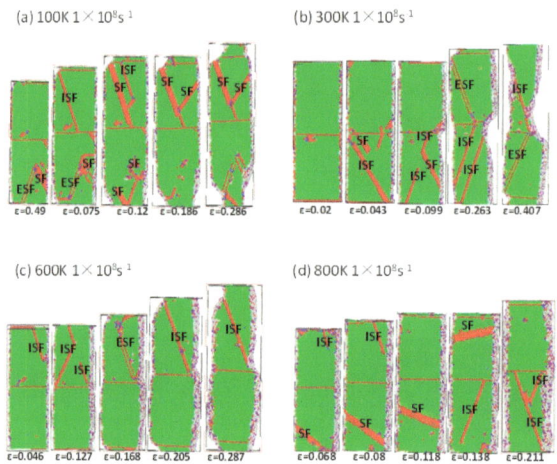

Figure 11. The CNA diagram of GBs HEAs under tensile. (**a**) 100 K, 1×10^8 s^{-1}; (**b**) 300 K, 1×10^8 s^{-1}; (**c**) 600 K, 1×10^8 s^{-1}; (**d**) 800 K, 1×10^8 s^{-1}.

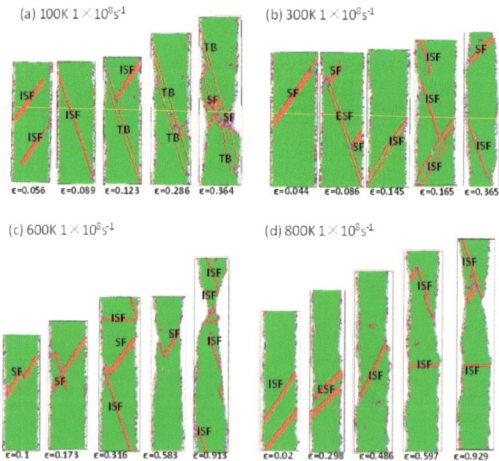

Figure 12. The CNA diagram of non-GBs HEAs under tensile. (**a**) 100 K, 1×10^8 s^{-1}; (**b**) 300 K, 1×10^8 s^{-1}; (**c**) 600 K, 1×10^8 s^{-1}; (**d**) 800 K, 1×10^8 s^{-1}.

The microstructure of dislocation lines Σ3 GBs HEAs (Figure 13) and non-GBs HEAs (Figure 14) at different temperatures (100, 300, 600, and 800 K) were researched. Dislocation type includes mainly green 1/6<112> Shockley partial dislocations, a small account of 1/6<110> Stair-rod, 1/2<110>perfect dislocation, and 1/3<111> Hirth dislocations. Draw a DXA diagram of Σ3 GBs HEAs at different temperatures (Figure 13), 1/6<112> Shockley partial dislocations always start nucleation near grain boundaries and form a dislocation loop. Dislocations expand with the increase in strain. With the increase in dislocation density, if it slips to the grain boundary, it is hard to slip and stack at grain boundaries. Therefore, dislocation entanglement and dislocation accumulation near grain boundaries. Figure 14 shows the motion of the dislocation line of the non-GBs HEAs at different temperatures. Dislocations appear randomly at the origin of non-GBs HEAs. The dislocation has been expanded smoothly in non-GBs HEAs. The dislocation density at 100 and 300 K is greater than 600 and 800 K. As the temperature increases, the movement of dislocations increases, and the annihilation of dislocations occurs, resulting in a decrease in dislocation density.

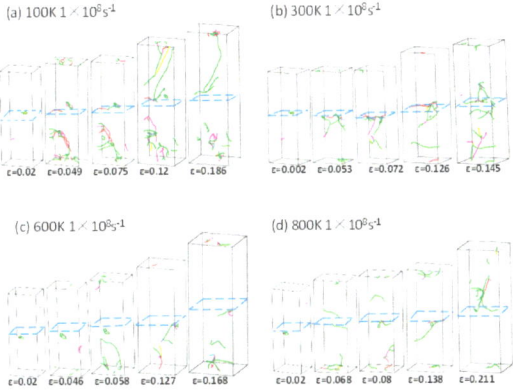

Figure 13. The DXA diagram of GBs HEAs under tensile. (a) 100 K, 1×10^8 s^{-1}; (b) 300 K, 1×10^8 s^{-1}; (c) 600 K, 1×10^8 s^{-1}; (d) 800 K, 1×10^8 s^{-1}.

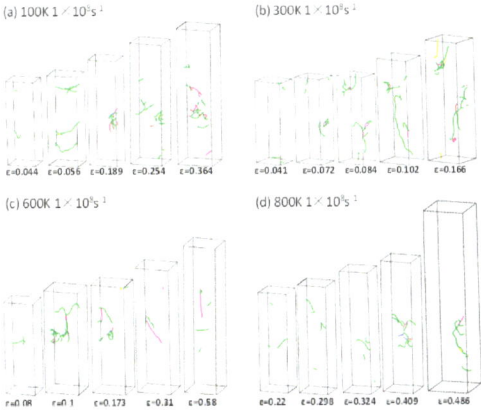

Figure 14. The DXA diagram of non-GBs HEAs under tensile. (a) 100 K, 1×10^8 s^{-1}; (b) 300 K, 1×10^8 s^{-1}; (c) 600 K, 1×10^8 s^{-1}; (d) 800 K, 1×10^8 s^{-1}.

4. Conclusions

The stress-strain curves of GBs HEAs and non-GBs HEAs at different strain rates and temperatures were compared and discussed. The variation trend for stacking faults and dislocations through the molecular dynamics method was researched.

The stress-strain curves of Σ3 GBs HEAs and non-GBs HEAs were analyzed at different strain rates at 300 K. From the elastic stage of approximate linear change to the peak yield strength, the system enters a long plastic deformation process until fracture. At the same strain rate, the yield strength of Σ3 GBs HEAs is higher than that of non-GBs HEAs. This is due to dislocation entanglement and dislocation accumulation near grain boundaries. There is no obvious difference in Young's modulus between Σ3 GBs HEAs and non-GBs HEAs. It indicates that the presence or absence of grain boundary has little effect on Young's modulus of HEAs. At different strain rates, with the increase in strain rate, the yield strength of Σ3 GBs HEAs and non-GBs HEAs increase with the increase in strain rate.

Under different temperatures, Σ3 GBs HEAs and non-GBs HEAs were subjected to uniaxial tensile at the same strain rate. The yield strength and Young's modulus decreased with the increase in temperature. The yield strength and Young's modulus of Σ3 GBs HEAs are higher than non-GBs HEAs at the same temperature. At the same time, with the increase in temperature, the dislocation of the Σ3 GBs HEAs and non-GBs HEAs are enhanced and occurred dislocation annihilation, resulting in the decrease in dislocation density. Dislocation type includes mainly 1/6<112> Shockley partial dislocations, a small account of 1/6<110> Stair-rod, 1/2<110>perfect dislocation, and 1/3< 111> Hirth dislocations. This research is helpful to understand how grain boundaries affect the mechanical properties of HEAs from the atomic point of view and provides guidance for the study of HEAs with better mechanical properties.

Author Contributions: C.L.: Methodology, Software, Validation, Investigation, Visualization, Writing—original draft, Writing—review and editing. R.W.: Software, Writing—review and editing. Z.J.: Methodology, Writing—reviewand editing. All authors have read and agreed to the published version of the manuscript.

Funding: This research was funded by the Natural Science Foundation of China (grant number 51971166) and the Key Laboratory of Shaanxi Provincial Education Department (grant number 20JS055).

Institutional Review Board Statement: Not applicable.

Informed Consent Statement: Not applicable.

Data Availability Statement: Data are contained within the article. The data presented in this study can be seen in the content above.

Acknowledgments: This work was supported by the Natural Science Foundation of China (grant number 51971166) and the Key Laboratory of Shaanxi Provincial Education Department (grant number 20JS055).

Conflicts of Interest: The authors declare no conflict of interest.

References

1. Chou, Y.-L.; Wang, Y.-C.; Yeh, J.-W.; Shih, H.-C. Pitting corrosion of the high-entropy alloy $Co_{1.5}CrFeNi_{1.5}Ti_{0.5}Mo_{0.1}$ in chloride-containing sulphate solutions. *Corros. Sci.* **2010**, *52*, 3481–3491. [CrossRef]
2. Shi, Y.; Yang, B.; Xie, X.; Brechtl, J.; Dahmen, K.A.; Liaw, P.K. Corrosion of AlxCoCrFeNi High-entropy alloys: Al-content and potential scan-rate dependent pitting behavior. *Corros. Sci.* **2017**, *119*, 33–45. [CrossRef]
3. Chuang, M.-H.; Tsai, M.-H.; Wang, W.-R.; Lin, S.-J.; Yeh, J.-W. Microstructure and wear behavior of $AlxCo_{1.5}CrFeNi_{1.5}Tiy$ high-entropy alloys. *Acta Mater.* **2011**, *59*, 6308–6317. [CrossRef]
4. Poulia, A.; Georgatis, E.; Lekatou, A. Microstructure and wear behavior of a refractory high entropy alloy. *Int. J. Refract. Met. Hard Mater.* **2016**, *57*, 50–63. [CrossRef]
5. Yeh, J.-W.; Chen, S.-K.; Lin, S.-J.; Gan, J.-Y.; Chin, T.-S.; Shun, T.-T.; Tsau, C.-H.; Chang, S.-Y. Nanostructured high-entropy alloys with multiple principal elements: Novel alloy design concepts and outcomes. *Adv. Eng. Mater.* **2004**, *5*, 299–303. [CrossRef]
6. Qiu, X.-W.; Zhang, Y.-P.; He, L.; Liu, C.-G. Microstructure and corrosion resistance of AlCrFeCuCo high entropy alloy. *J. Alloys Compd.* **2013**, *549*, 195–199. [CrossRef]
7. Yen, J.-W.; Lin, S.-J.; Chin, T.-S.; Gan, J.-Y.; Chen, S.-K.; Shun, T.-T.; Tsau, C.-H.; Chou, S.-Y. Formation of simple crystal structures in Cu-Co-Ni-Cr-Al-Fe-Ti-V alloys with multi principal metallic elements. *Metall. Mater. Trans. A* **2004**, *35*, 2533–2536.
8. Tong, C.-J.; Chen, Y.-L.; Yeh, J.-W.; Shun, T.-T.; Tsau, C.-H.; Lin, S.-J.; Chang, S.-Y. Microstructure characterization of AlxCoCr-CuFeNi high-entropy alloy system with multi-principal elements. *Metall. Mater. Trans. A* **2005**, *36*, 881–893. [CrossRef]

9. Wu, Z.; Bei, H.; Pharr, G.-M.; George, E.-P. Temperature dependence of the mechanical properties of equiatomic solid solution alloys with face-centered cubic crystal structures. *Acta Mater.* **2014**, *81*, 428–441. [CrossRef]
10. Liu, W.-H.; Wu, Y.; He, J.-Y.; Nieh, T.-G.; Lu, Z.P. Grain growth and the Hall–Petch relationship in a high-entropy FeCrNiCoMn alloy. *Scr. Mater.* **2013**, *68*, 526–529. [CrossRef]
11. Thota, H.; Jeyaraam, R.; Bairi, L.-R.; Tirunilaic, A.-S.; Kauffmann, A.; Freudenberger, J.; Heilmaier, M.; Mandal, S.; Vadlamani, S.-S. Grain boundary engineering and its implications on corrosion behavior of equiatomic CoCrFeMnNi high entropy alloy. *J. Alloys Compd.* **2020**, *888*, 161500. [CrossRef]
12. Chen, B.-R.; Yeh, A.-C.; Yeh, J.-W. Effect of one-step recrystallization on the grain boundary evolution of CoCrFeMnNi high entropy alloy and its subsystems. *Sci. Rep.* **2016**, *6*, 22306. [CrossRef]
13. Li, Z.-M.; Tasan, C.-C.; Pradeep, K.-G.; Raabe, D. A TRIP-assisted dual-phase high-entropy alloy: Grain size and phase fraction effects on deformation behavior. *Acta Mater.* **2017**, *131*, 323–335. [CrossRef]
14. Zhao, S.-J. Effects of local elemental ordering on defect-grain boundary interactions in high-entropy alloys. *J. Alloys Compd.* **2021**, *887*, 161314. [CrossRef]
15. Kumar, N.; Ying, Q.; Nie, X.; Mishra, R.-S.; Tang, Z.; Liaw, P.-K.; Brennan, R.-E.; Doherty, K.-J.; Cho, K.-C. High Strain-Rate Compressive Deformation Behavior of the $Al_{0.1}$CrFeCoNi High Entropy Alloy. *Mater. Des.* **2015**, *86*, 598–602. [CrossRef]
16. Bulatov, V.-V.; Reed, B.-W.; Kumar, M. Grain boundary energy function for fcc metals. *Acta Mater.* **2014**, *65*, 161–175. [CrossRef]
17. Yang, J.; Qiao, W.-J.; Ma, G.-S.; Wu, Y.-G.; Zhao, D. Revealing the Hall-Petch relationship of $Al_{0.1}$CoCrFeNi high-entropy alloy and its deformation mechanisms. *J. Alloys Compd.* **2019**, *795*, 269–274. [CrossRef]
18. Yang, T.; Xia, S.; Liu, S.; Wang, C.; Wang, L. Effects of Al addition on microstructure and mechanical properties of AlxCoCrFeNi High-entropy alloy. *Mater. Sci. Eng. A* **2015**, *648*, 15–22. [CrossRef]
19. Xia, S.; Zhang, Y. Deformation mechanisms of $Al_{0.1}$CoCrFeNi high entropy alloy at ambient and cryogenic temperatures. *Mater. Sci. Eng. A* **2018**, *733*, 408–413. [CrossRef]
20. Yang, T.-F.; Tang, Z.; Xie, X.; Carroll, R.; Wang, G.-Y.; Wang, Y.-G.; Dahmen, K.-A.; Liaw, P.-K.; Zhang, Y.-W. Deformation mechanisms of $Al_{0.1}$CoCrFeNi at elevated temperatures. *Mater. Sci. Eng. A* **2017**, *684*, 552–558. [CrossRef]
21. Jia, L.; Fang, Q.-H.; Liu, B.; Liu, Y.-W.; Liu, Y. Mechanical behaviors of AlCrFeCuNi high-entropy alloys under uniaxial tensile via molecular dynamics simulation. *RSC Adv.* **2016**, *6*, 76409–76419.
22. Brandon, D.-G. The structure of high-angle grain boundaries. *Acta Metall.* **1966**, *14*, 1479–1484. [CrossRef]
23. Yang, Y.-C.; Liu, C.-X.; Lin, C.-Y.; Xia, Z.-H. The effect of local atomic configuration in high-entropy alloys on the dislocation behaviors and mechanical properties. *Mater. Sci. Eng. A* **2021**, *815*, 141253. [CrossRef]
24. Sharma, A.; Balasubramanian, G. Dislocation dynamics in $Al_{0.1}$CoCrFeNi high-entropy alloy under tensile loading. *Intermetallics* **2017**, *91*, 31–34. [CrossRef]

Article

The Influence of Ion Beam Bombardment on the Properties of High Laser-Induced Damage Threshold HfO$_2$ Thin Films

Yingxue Xi *, Jiwu Zhao, Jin Zhang, Changming Zhang and Qi Wu

School of Optoelectronic Engineering, Xi'an Technological University, Xi'an 710021, China; zhaojiwu@st.xatu.edu.cn (J.Z.); j.zhang@xatu.edu.cn (J.Z.); zhangcm1995@163.com (C.Z.); 18829897910@163.com (Q.W.)
* Correspondence: xiyingxue@163.com

Abstract: HfO$_2$ thin films were deposited on BK-7 glass substrates using an electron beam evaporation deposition (EBD) technique and then post-treated with argon and oxygen ions at an ion energy ranging from 800 to 1200 eV. The optical properties, laser damage resistance, and surface morphology of the thin films exposed to Ar ions and O$_2$ ions at various energies were studied. It was found that the two ion post-treatment methods after deposition were effective for improving the LIDT of HfO$_2$ thin films, but the mechanism for the improvement differs. The dense thin films highly resistant to laser damage can be obtained through Ar ion post-treatment at a certain ion energy. The laser-induced damage threshold (LIDT) of thin films after O$_2$ ion post-treatment was higher in comparison to those irradiated with Ar ion at the same ion energy.

Keywords: HfO$_2$ films; the refractive index; laser-induced damage threshold; ion post-treatment

1. Introduction

Optical films with laser damage resistant properties are in considerable demand in high energy laser applications [1]. Hafnium dioxide (HfO$_2$) is one of the most important oxides materials with a high refractive index for the manufacture of interference multilayer films because of its excellent optical, thermal, and mechanical properties and is also known as a high laser damage threshold (LIDT) material [2]. It is well known that the laser damage resistance of HfO$_2$ thin film is dependent on the parameters of the manufacturing procedure. Segregation can be introduced during the manufacturing process of the films, making the laser damage resistance of the HfO$_2$ thin film deteriorate [3–5]. Additionally, dense morphology is also essential for a HfO$_2$ thin film with high laser damage threshold.

Ion beam processing techniques have the advantage in the surface and physical properties' modification of many optical films [6–8]. Irradiation through ionizing radiation has been developed to bring about changes in the oxide thin film's structure and physical properties. Such changes may be in the form of thermal absorption or stoichiometry of oxidation hinging upon both the chemical and physical nature of the films. As a matter of fact, thermal absorption can inflict damage on optical films when exposed to the radiation of high-power lasers [9–11].

Extensive research has been carried out into the effect of ion post-treatment on the intrinsic properties of HfO$_2$. Nevertheless, some reported results are contradictory, particularly those regarding ions and energy. Therefore, further clarifying studies are required. The current work comparatively studies various changes occurring in the intrinsic properties of HfO$_2$ films when exposed to argon and oxygen ion radiation at different energies, respectively.

2. Experimental Details

HfO$_2$ films were grown on BK-7 glass discs (diameter—25 mm) by means of electron-beam heating sources for the efficient evaporation of high purity particles. A standard RCA

Citation: Xi, Y.; Zhao, J.; Zhang, J.; Zhang, C.; Wu, Q. The Influence of Ion Beam Bombardment on the Properties of High Laser-Induced Damage Threshold HfO$_2$ Thin Films. *Crystals* **2022**, *12*, 117. https://doi.org/10.3390/cryst12010117

Academic Editors: Dah-Shyang Tsai and Mingjun Huang

Received: 24 November 2021
Accepted: 12 January 2022
Published: 17 January 2022

Publisher's Note: MDPI stays neutral with regard to jurisdictional claims in published maps and institutional affiliations.

Copyright: © 2022 by the authors. Licensee MDPI, Basel, Switzerland. This article is an open access article distributed under the terms and conditions of the Creative Commons Attribution (CC BY) license (https://creativecommons.org/licenses/by/4.0/).

cleaning process was implemented to clean the specimens before the process of deposition, then the specimens were immediately laid into the coating chamber.

The preparation experiments of HfO$_2$ films were performed using a ZZS500—2/G system from Chengdu Rankuum Machinery Limited, China. Prior to deposition. The vacuum pressure of the chamber was not less than 3.0 × 10^{-3} Pa, and the substrate temperature was maintained at 200 °C. High purity sintered HfO$_2$ pellets were evaporated at the rate of 12.5 nm/min, and the thickness of the films was monitored by the turning point monitoring approach of photoelectricity. To compensate for the loss of oxygen, high purity oxygen (99.999% in purity) was emptied into the chamber through separate mass flow controllers. After the deposition, a cold-cathode ion source was installed into the chamber for ion post-treatment. Ion irradiation was carried out at various ion energies from 800 to 1200 eV. The ion flux was 20 µA/cm^2, and the treatment duration was 15 min. To identify the influence of ion species, the ion beams used to bombard the surface of the as-deposited films samples discharged high purity argon and oxygen gases, respectively.

In order to investigate changes in the optical properties of irradiated HfO$_2$ thin films, the refractive index was determined by using a M-2000UI spectroscopic ellipsometer (SE) manufactured by J.A.Woollam company of United States. Measured data were used to describe an optical model that can help to gather the thickness and optical properties by conducting regression analysis. The HfO$_2$ film on glass substrate was initially assumed to have a four phase system (from top to bottom): the incident medium (air), the roughness layer, the HfO$_2$ layer, and the glass substrate, as shown in Figure 1 (insert). An empirical formula of Cauchy's model was applied to calculate n(λ), which was given by adjusting the fitting parameters according to the SE measured data of the thin HfO$_2$ film due to its weak absorbance of light in the 400~900 nm wavelength range (i.e., k is negligible).

$$n(\lambda) = A + \frac{B}{\lambda^2} + \frac{C}{\lambda^3} \qquad (1)$$

where λ is the wavelength of incident light, and A, B, and C are empirical constants. The typical experimental data, model fit to the data, and the fitted parameters are shown in Figure 1. The mean square error (MSE) of all samples in our experiment was less than 10, which indicates the measured and calculated results were in marked correspondence.

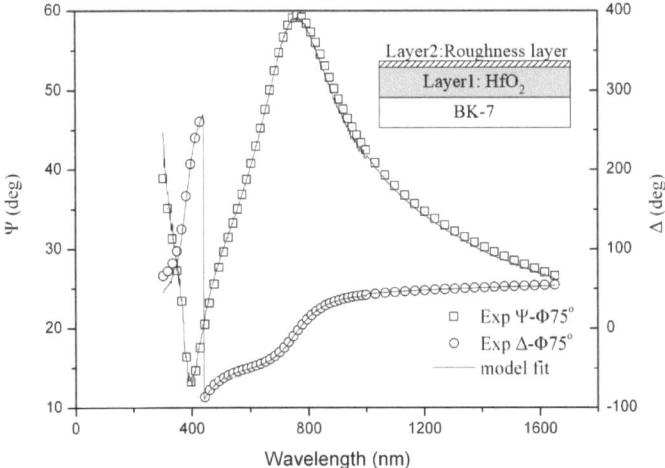

Figure 1. Typical HfO$_2$ thin film measured data and fitting curves with fits to the Cauchy formula. The optical model is shown in the inserts at the top right corner.

An LIDT test at 1064 nm was performed in the 1-on-1 mode in the light of ISO standard 11254-1. A new in situ laser damage image test apparatus was set up, as shown in Figure 2. The samples were exposed to a laser beam with 1064 nm wavelength and 12-ns effective pulse duration for the Nd: YAG laser system. The pulse energy was adjusted with an optical attenuator in a given fluence range and monitored in realtime with a laser energy meter. The morphology of laser damage on the surface of samples was investigated by means of a CCD camera-microscope device (magnification ×100), which ensured realtime testing and recording of the irradiated zone in situ [12].

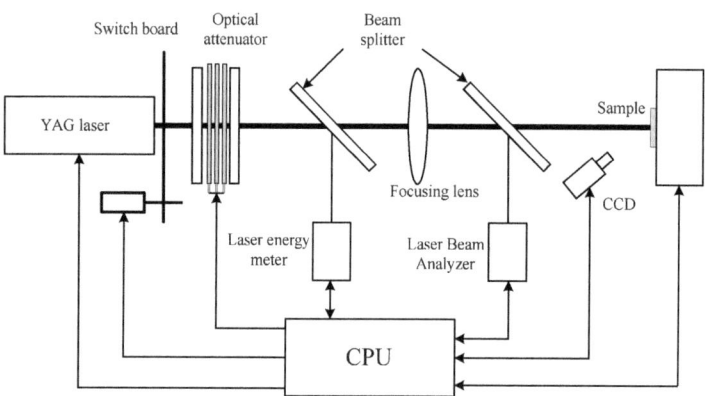

Figure 2. Schematic diagram of the apparatus used for in situ image laser damage testing system.

The laser light was focused down to the small spot size of 0.8 mm onto the film sample surface. Using this device, the probability curves of damage were expressed, in the meantime, via tallying up the quantity of damaged regions at every fluence F, the damage probability P(F) was estimated. This test was performed via testing 100 points for laser radiation. The images of the test site before and after each shot was observed for a certain energy, which was divided into 10 different levels to identify a high-accuracy LIDT of the sample.

The films' surface roughness was measured using a Talysurf CCI 2000 non-contact 3D profiler (Taylor Hobson Limited, Leicester, UK), and the root mean square roughness value (RMS) was presented.

3. Results and Discussion

X-ray Photoelectron Spectroscopy (XPS) is one of the important ways to study the electronic and atomic structure of materials. In this experiment, the thin film samples were analyzed by a PHI-5400 X-ray photoelectron spectroscopy from the American PE company, and Cu-Kα radiations(= 1.54 Å) were used as the X-ray source. Before depositions, all the film sample surfaces were subjected to argon ion etching for 90 s, in which the energy value of argon ion bombardment was 2 keV. In addition, the XPS spectra in this experiment were corrected by the binding energy of C1s orbital 284.8 eV. To obtain the accurate data of Hf and O elements in the film, the two elements were scanned by a fine spectrum, as shown in Figure 3.

According to the peak area of Hf4f and O1s, the stoichiometric ratio of Hf and O can be determined, and the expression is:

$$N_1 : N_2 = A_1 S_2 / A_2 S_1 \tag{2}$$

In the above formula, $N_1:N_2$ is the stoichiometric ratio of elements Hf and O; A_1 and A_2 are the spectral peak areas corresponding to the elements ($A_1 = 270{,}580$ and $A_2 = 155{,}441$ here); and S_1 and S_2 are sensitivity factors corresponding to elements [13] (here $S_1 = 0.71$ and $S_2 = 2.221$), The stoichiometric ratio of Hf and O is 1:1.78. After oxygen ion bombardment at ion energy of 1000 eV, the ratio can be increased to 1:1.86 ($A_1 = 257{,}609$ and $A_2 = 154{,}454$

here). The stoichiometric ratio of Hf and O irradiated film using an O Ion beam are closer to 1:2 than the as-deposited films without irradiation.

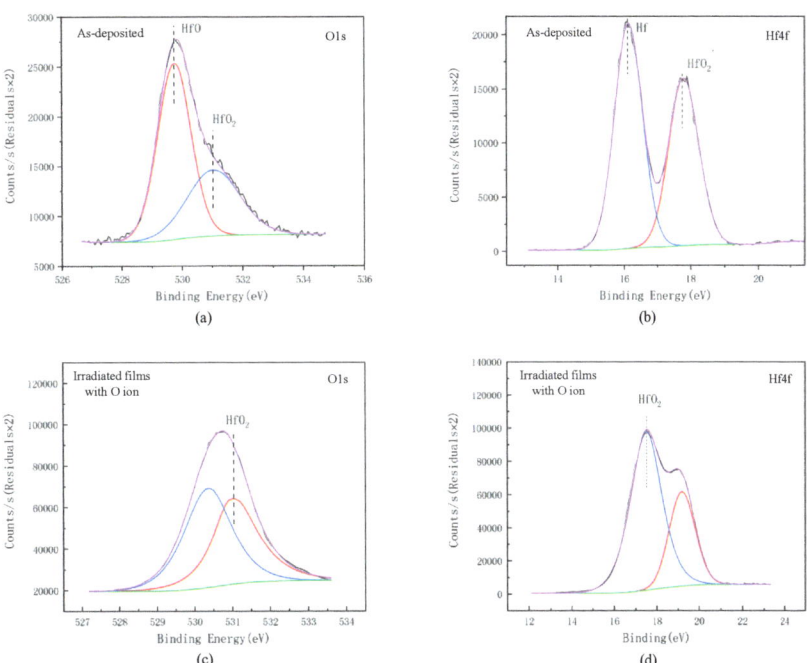

Figure 3. Comparison of XPS patterns of HfO$_2$ films before and after O ions treatment at an ion energy of 1000 eV; (**a**) O1s spectra of as-deposited films; (**b**) Hf4f spectra of as-deposited films; (**c**) O1s spectra of irradiated films; and (**d**) Hf4f s spectra of irradiated films.

The crystal structure of HfO$_2$ film was measured by Brooke-D2 X-ray diffractometer (XRD). During the test, the scanning step was 0.02°, the scanning area was 20°~100°, and Figure 4 shows the test results. It can be observed from the figure that a typical monoclinic HfO$_2$ characteristic peak appeared in the diffraction pattern of the prepared film and had a preferred orientation in the (311) direction. The test results show that crystallization occurred in the HfO$_2$ films deposited via electron beam evaporation deposition technology; moreover, the prepared films were polycrystalline films [14].

The changes in the optical properties of irradiated HfO$_2$ films were analyzed by SE. Figure 5a,b, respectively, show the dispersion of the refractive index of HfO$_2$ newly deposited film and irradiated films exposed to ion beam irradiation at different ion energies under Ar and O$_2$ atmosphere. From Figure 5, it can be seen clearly that the refractive index for all samples investigated sharply increased with the wavelength between 400 and 450 nm, this anomalous dispersion is speculated to be caused by absorption in a shorter wavelength region for HfO$_2$ thin films. When the energy of irradiation ion beam varied between 800 and 1200 eV, the refractive index of the irradiated films had significant and obvious changes in the given range of measurement wavelength, which was affected by the species of ion beam. It could also be observed that the n value increased along with the increase in the irradiation Ar ion energy; whereas, in comparison with the as-deposited films, the value of the refractive index decreased when it was beyond 1000 eV.

Many porous and void-rich structures were formed easily in the HfO$_2$ thin film preparation by EBD. However, as shown in Figure 5a, the ion bombardment effect of argon plasma-treated at different energies from 800 to 1200 eV for as-deposited samples could make films become denser and solid, and more apparent ion bombardment effect in the aspect of optical properties was observed with the improvement of ion energy. On the other hand, oxygen was easily emitted from HfO_2 during the deposition process, which was an imperative factor influencing the optical properties; in addition, hafnium was absorbed largely in the UV band, which increased the n value to a certain extent [15]. From Figure 5b, the n value was slightly lowered by ion bombardment with oxygen plasma. It obviously reveals that low energy oxygen ions could improve the stoichiometry in thin films.

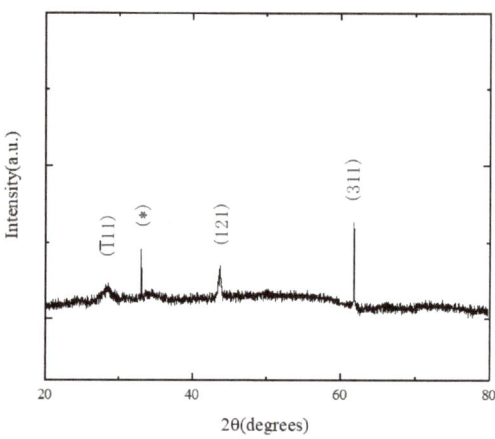

Figure 4. XRD phase diagram of HfO_2 film; * indicates the SiO_2 (217) peak.

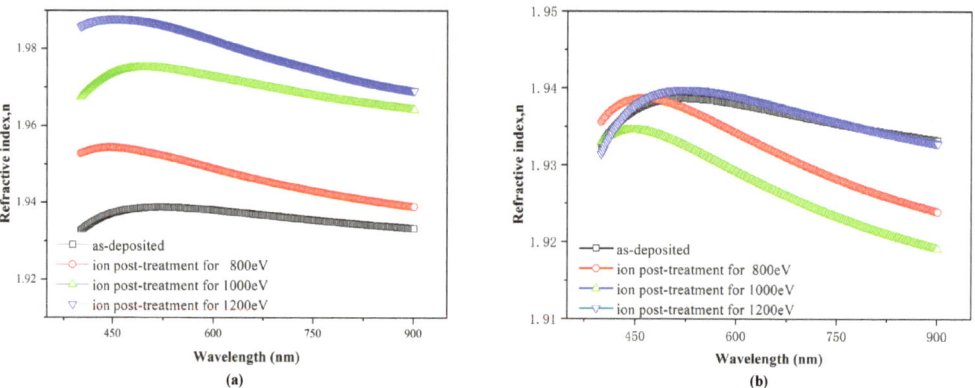

Figure 5. Refractive index of HfO_2 film irradiated at various ion energies under Ar (**a**) and O_2 (**b**) atmosphere.

The effect of ion energy and species on LIDT of HfO_2 thin films in the ion post-treatment are shown in Figure 6. The LIDT in the deposited HfO_2 thin film was found by 16.31 J/cm², but the maximum thresholds of 19.7 J/cm² and 25.72 J/cm² were observed in the irritated films at 800 eV under Ar and O_2, respectively. This shows ion post-treatment had a significant contribution on the laser damage resistance. As shown in Figure 5a, the thin film became denser after post-treatment with Ar^+ plasma. A plasma-treated ion beam can lessen the quantity of thermal defects for irradiated HfO_2 thin film, whose thermal conductivity was higher than that of the as-deposited samples [16,17]. Furthermore, ion treatment is also likely to be a feasible method of making the defects in the coatings stabilized to avoid laser damage. However, the LITD decreased gradually with the further increase in the ion energy, and the LIDT of HfO_2 irradiated at 1200 eV was even lower than that of the as-deposited, whose values were only 12.21 J/cm² for Ar^+ plasma and

15.14 J/cm² for O_2^+ plasma. A conceivable explanation for the decline in the LIDT value of the irradiated samples is that irradiation by high energy ions results in an increase in internal defects to reduce the laser damage resistance.

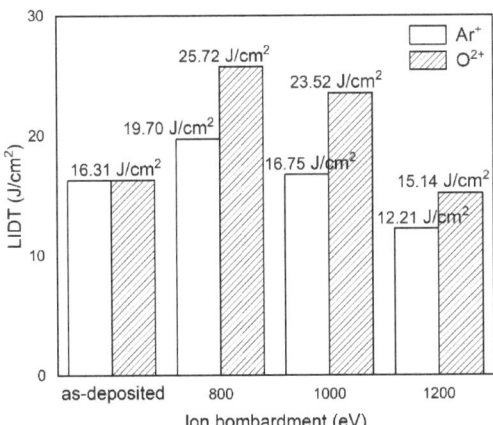

Figure 6. Laser damage threshold of HfO_2 film irradiated at various ion energies under Ar and O_2 atmosphere.

Although changes in the LIDT with the increase in the energy were similar for ion post-treatment with Ar^+ and O_2^+ plasma, the films irritated by O_2^+ plasma had higher LITD at the same ion energy. The reduction in the substoichiometric ratio oxygen to hafnium on irradiation might have a bearing on this difference [18,19], because the decrease in the sub-oxide component of HfO_2 films during irradiation was correlated with the O_2 ion bombardment. In fact, the two most essential factors that bear on the LIDT of thin film were defect density and absorption. The theory of electron-avalanche-ionization can be used to expound and explicate the effect of substoichiometer compositions on LIDT of HfO_2 thin films. Ion beam bombardment of oxygen plasma could repair oxygen vacancies and reduce the absorption of the as-deposited films. The oxygen post-treatment on the anti-reflective film reported by Yuan [20] is in good agreement with these results. However, they performed a direct contrast with those reported for HfO_2 films irradiated using end-hall ion source, which were found to be little changed in the LITD on low energy oxygen ions.

Figure 7 shows the graph of the RMS roughness values of irradiated films for different ion bombardment energies. A similar increasing trend of RMS value with an increase in ion bombardment energy for each film irradiated with Ar and O_2 plasma was observed, but the increase in each change is not alike. As for newly deposited films, the minimum roughness value was 2.8 nm, which indicated an atomically smooth surface. With an increase from 800 to 1200 eV in the ion energy, the surface roughness for post-treated film in O_2 increased gradually from 3.9 to 18.3 nm, but the surface films post-treated in Ar were rougher, increasing from 6.6 to 24.5 nm. This may be due to the larger mass of Ar than O_2 [21,22]. The thermal spike phenomenon caused by ion-bombardment is the main reason leading to an increase in the surface roughness of irradiated film.

For irradiated film, the surface roughness increased proportionately with ion energy but inversely with the LIDT value. It seems that the LIDT value might be related to the change in the surface roughness induced by ion bombardment [23], especially for films irradiated with high energetic plasma ions. Deeper and further analysis is still required for more fundamental comprehension.

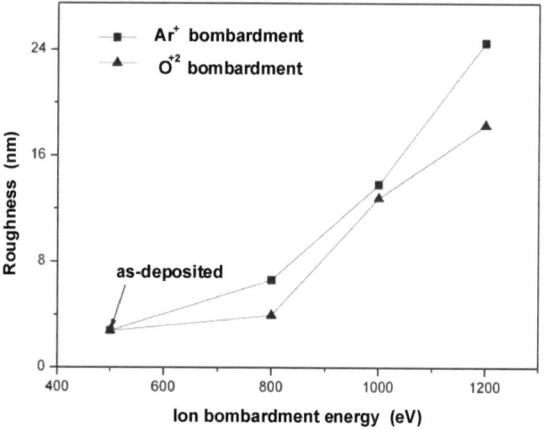

Figure 7. Variation of roughness of HfO_2 films with Ar and O_2 ion energy.

4. Conclusions

HfO_2 thin films were processed in advance by the EBE technique at 200 °C and then subjected to ion post-treatment in Ar and O_2 plasma at various ion energies from 800 to 1200 eV, respectively. The influence of the ion energy and species on the optical properties, laser damage resistance, and surface morphology were systematically studied. The refractive index of the film increased with ion energy in the range of 800~1200 eV after argon ion post-treatment; however, the refractive index of thin films irradiated in oxygen plasma decreased with the increase in the ion energy except 1200 eV, and it was inclined to show a very minor shift towards lower values that rested on ion energy. The laser damage resistance of HfO_2 was strongly dependent on ion irradiation energy and ion species. The LIDT of HfO_2 films irradiated at certain ion energies were improved but decreased with the increase in ion energy up to 1000 eV. Moreover, O_2 ion irradiation was better than argon ion as a means of improving LITD in thin film. The LITD in the films after ion post-treatment was inversely proportional to the surface roughness. A conclusion can be drawn that the laser damage properties of irradiated HfO_2 thin film may be related to the change in the surface roughness induced by ion bombardment.

Author Contributions: Conceptualization, Y.X. and J.Z. (Jiwu Zhao); Data curation, J.Z. (Jiwu Zhao) and C.Z.; Formal analysis, J.Z. (Jiwu Zhao); Funding acquisition, Y.X.; Investigation, Y.X. and J.Z. (Jiwu Zhao); Methodology, J.Z. (Jin Zhang); Project administration, Y.X. and J.Z. (Jiwu Zhao); Resources, Y.X. and J.Z. (Jiwu Zhao); Supervision, J.Z. (Jiwu Zhao); Validation, J.Z. (Jin Zhang); Visualization, Q.W.; Writing—original draft, Y.X. All authors have read and agreed to the published version of the manuscript.

Funding: This study was financially supported by the intergovernmental international scientific and technological innova-tion cooperation of science and technology ministry of China (No.2018YFE0199200), This paper also been supported by the Science and Technology on Applied Physical Chemistry Laboratory.

Institutional Review Board Statement: Not applicable.

Informed Consent Statement: Not applicable.

Data Availability Statement: Not applicable.

Conflicts of Interest: The authors declare no conflict of interest.

References

1. Zhao, Z.; Sun, J.; Zhu, M. Research to improve the optical performance and laser-induced damage threshold of hafnium oxide/silica dichroic coatings. *Opt. Mater.* **2021**, *113*, 110890. [CrossRef]
2. Field, E.S.; Galloway, B.R.; Kletecka, D.E. Dual-wavelength laser-induced damage threshold of a HfO_2/SiO_2 dichroic coating developed for high transmission at 527 nm and high reflection at 1054 nm. *Proc. SPIE Laser Damage.* **2019**, *11173*, 1117314.
3. Papernov, S.; Kozlov, A.A.; Oliver, J.B. Near-ultraviolet absorption annealing in hafnium oxide thin films subjected to continuous-wave laser radiation. *Opt. Engineering.* **2014**, *53*, 122504. [CrossRef]
4. Fang, M.H.; Tian, P.Y.; Zhu, M.D. Laser-induced damage threshold in HfO_2/SiO_2 multilayer films irradiated by β-ray. *Chin. Phys. B* **2019**, *28*, 024215. [CrossRef]
5. Dong, J.; Fan, J.; Mao, S. Effect of annealing on the damage threshold and optical properties of $HfO_2/Ta_2O_5/SiO_2$ high-reflection film. *Chin. Opt. Lett.* **2019**, *17*, 113101. [CrossRef]
6. Pan, F.; Wang, J.; Liu, M.; Wei, Y.; Liu, Z.; Zhang, F.; Wang, Z.; Luo, J.; Wu, Q.; Li, S. Influence of ion assistance on optical properties, residual stress and laser induced damage threshold of HfO_2 thin film by use of different ion sources. *Proc. SPIE Optifab.* **2019**, *11175*, 1117510.
7. Balogh-Michels, Z.; Stevanovic, I.; Borzi, A. Crystallization behavior of ion beam sputtered HfO_2 thin films and its effect on the laser-induced damage threshold. *J. Eur. Opt. Soc. Rapid Publ.* **2021**, *17*, 1–8. [CrossRef]
8. Pan, F.; Wei, Y.; Zhang, F. Correlation between the structure and laser damage properties of ion assisted HfO_2 thin films. *Int. Soc. Opt. Photonics.* **2019**, *11064*, 110640L.
9. Mao, S.; Fan, J.; Zou, Y. Effect of two-step post-treatment on optical properties, microstructure, and nanosecond laser damage threshold of $HfO_2/TiO_2/SiO_2$ multilayer high reflection films. *J. Vac. Sci. Technol. A Vac. Surf. Film.* **2019**, *37*, 061503. [CrossRef]
10. Wang, Y.; Ma, Y.; Wang, D. Theoretical simulation analysis of long-pulse laser induced damage in a BK7: SiO_2/HfO_2 optical anti-reflective films. *Optik* **2018**, *156*, 530–535. [CrossRef]
11. Liu, J.; Ling, X.; Liu, X. Mechanism of annealing effect on damage threshold enhancement of HfO_2 films in vacuum. *Vac.* **2021**, *189*, 110266. [CrossRef]
12. Negres, R.A.; Carr, C.W.; Laurence, T.A. Laser-induced damage of intrinsic and extrinsic defects by picosecond pulses on multilayer dielectric coatings for petawatt-class lasers. *Opt. Engineering.* **2016**, *56*, 011008. [CrossRef]
13. Liu, J.; Li, X.; Yu, Z. Effect of laser conditioning on the LIDT of 532 nm HfO_2/SiO_2 thin films reflectors. *Proc. SPIE/SIOM Pac. Rim Laser Damage.* **2013**, *8786*, 87860Z.
14. Alvisi, M.; De Tomasi, F.; Perrone, M.R. Laser damage dependence on structural and optical properties of ion-assisted HfO2 thin films. *Thin Solid Film.* **2001**, *396*, 44–52. [CrossRef]
15. Zhang, D.; Zhu, M.; Li, Y. Laser-induced damage of 355 nm high-reflective mirror caused by nanoscale defect. *J. Wuhan Univ. Technol. Mater. Sci. Ed.* **2017**, *32*, 1057–1060. [CrossRef]
16. Li, C.; Zhao, Y.; Cui, Y. Investigation on picosecond laser-induced damage in HfO_2/SiO_2 high-reflective coatings. *Opt. Laser Technol.* **2018**, *106*, 372–377. [CrossRef]
17. Jena, S.; Tokas, R.B.; Rao, K.D. Influence of oxygen partial pressure on microstructure, optical properties, residual stress and laser induced damage threshold of amorphous HfO_2 thin films. *J. Alloys Compd.* **2019**, *771*, 373–381. [CrossRef]
18. Xu, Y.; Dunlap, D.H.; Emmert, L.A. Laser-driven detonation wave in hafnium oxide film: Defect controlled laser damage and ablation. *J. Appl. Phys.* **2020**, *128*, 123101. [CrossRef]
19. Wei, Y.; Xu, Q.; Wang, Z. Growth properties and optical properties for HfO_2 thin films deposited by atomic layer deposition. *J. Alloys Compd.* **2018**, *735*, 1422–1426. [CrossRef]
20. Yuan, H.; Zhang, G.; Kan, S. Influence of the deposition method on the laser induced damage threshold of HfO_2 thin films. *J. Huazhong Univ. Sci. Technol. (Nat. Sci. Ed.)* **2007**, *35*, 108–111.
21. Wang, C.; Jin, Y.; Zhang, D. A comparative study of the influence of different post-treatment methods on the properties of HfO_2 single layers. *Opt. Laser Technol.* **2009**, *41*, 570–5736. [CrossRef]
22. Jena, S.; Tokas, R.B.; Rao, K.D. Annealing effects on microstructure and laser-induced damage threshold of HfO_2/SiO_2 multilayer mirrors. *Appl. Opt.* **2016**, *55*, 6108–6114. [CrossRef] [PubMed]
23. Kozlov, A.A.; Papernov, S.; Oliver, J.B. Study of the picosecond laser damage in HfO_2/SiO_2 based thin-film coatings in vacuum. *Proc. SPIE Laser Damage.* **2017**, *10014*, 100141Y.

Article

Effects of Sintering Processes on Microstructure Evolution, Crystallite, and Grain Growth of MoO₂ Powder

Jongbeom Lee [1,*], Jinyoung Jeong [1,2], Hyowon Lee [3], Jaesoung Park [3], Jinman Jang [1] and Haguk Jeong [1]

[1] Industrial Materials Processing R&D Group, Korea Institute of Industrial Technology, Incheon 21999, Republic of Korea; zero5677@kitech.re.kr (J.J.); man4502@kitech.re.kr (J.J.); hgjeong@kitech.re.kr (H.J.)
[2] Department of Advanced Materials Science and Engineering, Inha University, Incheon 22212, Republic of Korea
[3] Thin Film Materials R&D Team, LT Metal. Co., Ltd., Incheon 21697, Republic of Korea; hwlee@ltmetal.co.kr (H.L.); jaspark@ltmetal.co.kr (J.P.)
* Correspondence: ljb01@kitech.re.kr; Tel.: +82-32-850-0378

Abstract: MoO$_2$ micro-powders with a mean pore size of 3.4 nm and specific surface area of 2.5 g/cm^3 were compacted by dry pressing, then pressureless sintered at a temperature of 1000–1150 °C for 2 h or for a sintering time of 0.5–12 h at 1050 °C in an N$_2$ atmosphere. Then, their microstructure evolution for morphology, crystallite, and grain growth were investigated. By sintering at a certain temperature and times, the irregular shape of the MoO$_2$ powders transformed into an equiaxed structure, owing to the surface energy, which contributed to faster grain growth at the initial stage of sintering. The crystallite and grain sizes exponentially increased with the sintering time, and the growth exponent, n, was approximately 2.8 and 4, respectively. This indicates that the crystallite growth is governed by dislocation-mediated lattice diffusion, and the grain growth is determined by surface diffusion-controlled pore mobility. The increase in sintering temperature increased both crystallite and grain size, which obeyed the Arrhenius equation, and the activation energies were determined to be 95.65 and 76.95 kJmol^{-1} for crystallite and grain growths, respectively.

Keywords: MoO$_2$; sintering; XRD; SEM; morphology; crystallite; grain growth

Citation: Lee, J.; Jeong, J.; Lee, H.; Park, J.; Jang, J.; Jeong, H. Effects of Sintering Processes on Microstructure Evolution, Crystallite, and Grain Growth of MoO$_2$ Powder. *Crystals* **2023**, *13*, 1311. https://doi.org/10.3390/cryst13091311

Academic Editor: Andreas Thissen

Received: 5 August 2023
Revised: 23 August 2023
Accepted: 26 August 2023
Published: 28 August 2023

Copyright: © 2023 by the authors. Licensee MDPI, Basel, Switzerland. This article is an open access article distributed under the terms and conditions of the Creative Commons Attribution (CC BY) license (https://creativecommons.org/licenses/by/4.0/).

1. Introduction

Molybdenum oxides, a type of metal oxide with an *n*-type semiconducting and non-toxic nature, have attracted much attention for their diverse functional applications such as electronics, catalysis, sensors, energy-storage units, field emission devices, superconductor lubricants, thermal materials, biosystems, and chromogenic and electrochromic systems [1–3]. MoO$_X$ (2 ≤ X ≤ 3) has a high work function of approximately 6.7 eV and can be used as the hole-extraction layer in photovoltaic devices, light-emitting devices, and sensors, owing to its photochromism and low reflectance characteristics [4,5].

MoO$_X$ not only has many intermediate oxides such as MonO$_{3n-1}$ between MoO$_2$ and MoO$_3$, but also has various structural and electronic phases due to the multiple valence states of 4d molybdenum from +3 to +6 [6,7]. Among them, the most common are MoO$_3$ and MoO$_2$, which differ in their electronic and optical properties and chemical structures. MoO$_3$ has an orthorhombic structure and consists of two layers of MoO$_6$ octahedra as a framework [8]. MoO$_2$ has a disfigured rutile structure, which consists of MoO$_6$ octahedra coupled by edge-sharing [9]. Regarding the optical properties, MoO$_3$ usually presents as white, whereas MoO$_2$ exhibits as dark blue or black, allowing a different light absorption characteristic [10]. MoO$_2$ is electrically conductive, but MoO$_3$ is insulating [11,12]. This metallic property of MoO$_2$ endows it with possible applications as electrodes for gas sensors and catalysis [13–15].

MoO$_2$ is less important in technological applications than MoO$_3$ but has been used as a catalyst for alkane isomerization or oxidation reactions and as a gas sensor. Furthermore, MoO$_2$ is an oxide-based compound semiconductor with an indirect band gap, and it is also a potential candidate for desirable hole injection layer applications between a transparent conducting oxide (TCO) and an organic light-emitting diode (OLED) [16]. For the fine bezel in a thin-film transistor (TFT) backplane, MoO$_2$ can be applied as a functional film to lower the reflection of a metal-mesh electrode and enhance the effect of shadow elimination; additionally, it is a cost-effective and reliable thin-film material [17].

MoO$_2$ thin films can be fabricated using a molybdenum target using spin coating, pulsed laser deposition, reactive sputtering, and thermal evaporation [18]. In the case of the sputtering process, to optimize the performances of the thin film, most researchers have studied the influences of sputtering parameters on the film properties, including the type of sputter power (radio-frequency (RF) magnetron reactive sputtering, direct current (DC) magnetron reactive sputtering), working pressure (oxygen partial pressure from 1.00×10^{-3} mbar to 1.37×10^{-3} mbar), base pressure, atmosphere (Ar-O$_2$ sputtering gas), sputter powder (electrical conductivity varying from 1.6×10^{-5} S/cm to 3.22 S/cm), film thickness, and post-annealing treatment (oxygen annealing range from 250 to 350 °C) [19–23]. However, it is difficult to elucidate how the target influences the properties of various films, the sputtering process, and the correlations between film properties and target performance have not been well established. Therefore, recently, sputtering targets have gradually grown in importance because the sintered density, grain size, electrical properties, stoichiometry, and microstructural uniformity of the sputtering targets are now known to significantly influence not only the properties of various thin films but also the sputtering process. Understanding and controlling a sputtering target's crystallite and grain growth is important in thin film deposition processes, as it can affect the quality and properties of the deposited thin film [24].

The use of MoO$_2$ materials instead of Mo materials makes the optimization of process parameters easier because MoO$_2$ thin film is fabricated by reactive sputtering of Mo target materials in a controlled O$_2$ atmosphere. Several works [25–27] have investigated the fabrications and microstructural evolution of Mo materials by powder metallurgy, but studies on a fabrication of bulk materials using MoO$_2$ powders by means of the sintering process and its microstructural evolution during firing are limited. Therefore, this study systematically investigated commercially available MoO$_2$ powders to determine the effects of the sintering temperature and time on microstructure evolution for the morphology of the powder, grain growth. MoO$_2$ bulk materials were fabricated from MoO$_2$ powder using a conventional sintering process at different sintering temperatures and times. The crystalline structure, powder morphology, and microstructural evolution of the sintered MoO$_2$ were characterized via X-ray diffraction (XRD) and scanning electron microscopy (SEM). Moreover, the kinetics and activation energy for crystallite and grain growths were determined and elucidated based on lattice and surface diffusion mechanisms, respectively.

2. Materials and Methods

The as-received MoO$_2$ powders had irregular polygonal structures (Figure 1a), particle sizes with a mean diameter of 2.1 µm (Figure 1b), and a purity of 99.9%. The powders were compacted into a cylindrical shape with a diameter of 6 mm and thickness of 3 mm at a uniaxial press by applying 200 MPa at ambient temperature, and after pressure application the time of 10 min is provided for even spreading of pressure. The compacted samples were dried at 120 °C in an oven for 2 h before sintering. The MoO$_2$ sample was sintered at 1000–1150 °C for 2 h, and at 1050 °C for 0–12 h in an N$_2$ atmosphere. A constant heating rate of 5 °C/min and a dwell time of 2 h was applied at various temperatures. To assure specific temperature control, a type-B thermocouple (Thermo Fisher Scientific, Waltham, MA, USA) was applied in the furnace equipment, and the electric power obligated for the set sintering schedule was traced down using the potentiometer.

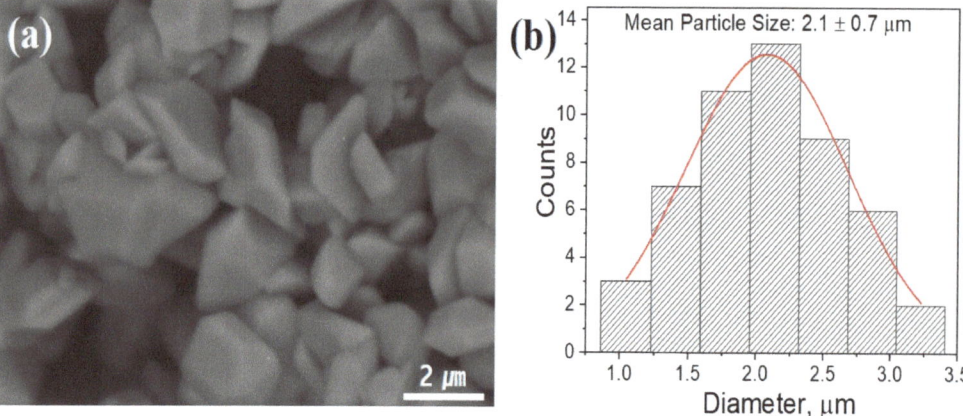

Figure 1. (a) SEM photo and (b) histogram demonstrating the particle size distribution of the as-received MoO_2 powders.

The specific surface area (SSA) and pore size of the MoO_2 micro-powders were determined using an automatic adsorption instrument (Quanta chrome Corp. Quadrasorb evo, Boynton Beach, FL, USA). Sample degassing was performed at 170 °C for 13 h, before the absorption and desorption of liquid N_2 at −196 °C (77 K). The MicroActive 4.0 software (TriStar II 3020 version 2.0) was used to generate the Brunauer–Emmett–Teller (BET) surface area and the Barrett–Joyner–Helenda (BJH) pore size distribution. The sintered samples were identified using X-ray diffractometry (XRD, Smartlab, Rigaku Co.,Tokyo, Japan) to determine the lattice parameters and dislocation density (Cu Kα, λ = 1.54059 Å, 25 mA × 40 kV power, in the range of 2θ = 20–80°, and a step size of 0.02°) at room temperature. The XRD results were obtained using the software for the powerful Rietveld refinement method (Crystal Impact GbR, Bonn, Germany). Highscore Plus files from CrystalMaker software version 10.8.2. were used to generate a crystal structure of the sintered sample. The morphology and microstructure of the powders and the fractured surface of the sintered samples were investigated by SEM at an accelerating voltage of 20 kV (SU5000, Hitachi, Tokyo, Japan). To impede the specimen charging, a layer of platinum was coated onto the powder and sintered samples for 15 s. The particle and grain sizes of the powders and sintered samples were measured using image analysis software (Image-Pro, Media Cybernetics Inc., Rockville, MD, USA).

3. Results and Discussion

The BET isotherms for nitrogen adsorption by the as-received MoO_2 powders were measured to evaluate the pore structure. As shown in Figure 2a, the distinctive hysteresis loop was mainly observed at a higher pressure, i.e., P/P_0 = 0.0–1.0, indicating a type IV isotherm, which is indicative of a typical mesoporous material [28]. The hysteresis loop is formed because of the difference in the capillary action in the mesopores and macropores during the absorption and desorption process [29]. The pore size distribution, determined from the adsorption–desorption curve using the BJH method, was analyzed as a function of the pore diameter of the MoO_2 powders, as shown in Figure 2b, which shows the pore size distribution for pore diameters ranging from 1 to 300 nm. The SSA of the powders determined from the N_2 adsorption–desorption isotherm using the BET method was 2.5 m^2g^{-1}, macropores of size larger than 50 nm rarely existed, and the mean pore size was approximately 3.4 nm (Figure 2b).

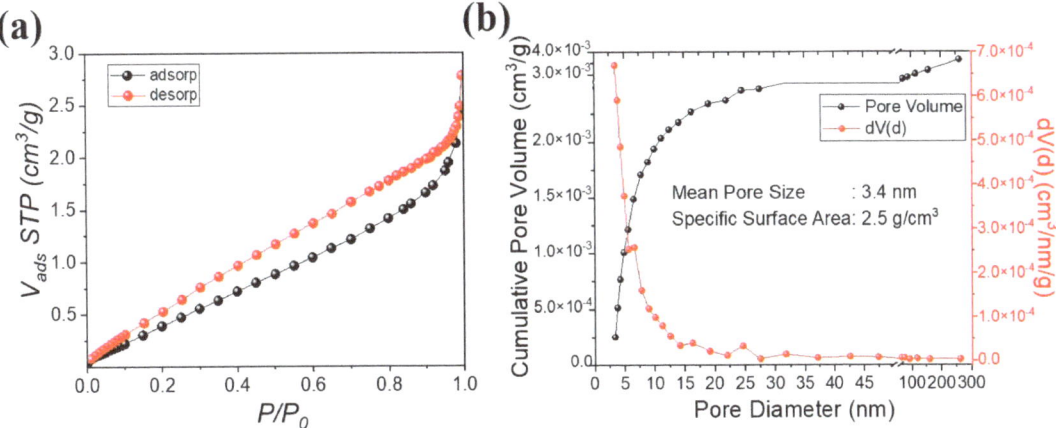

Figure 2. (**a**) N$_2$ adsorption-desorption isotherms and (**b**) cumulative pore volume and BJH pore size distribution curves of as-received MoO$_2$ powders.

XRD patterns and crystal structure modeling of the as-received and sintered samples are shown in Figure 3. The XRD results exhibit characteristic peaks, similar to a previous study [30], and the as-received powders have a monoclinic structure with a lattice parameter of a = 5.6102 Å, b = 4.8573 Å, c = 5.6265 Å, and β = 120.915°, showing crystal structure modeling (Figure 3b). The dashed lines in Figure 3a indicate the peak of the main *hkl* miller indices of the as-received and sintered samples, indicating that each lattice parameter is invariant under the sintering process in this study, while the phase transition of MoO$_2$ from monoclinic structure to a tetragonal structure at T = 1260 °C (1533 K) was reported in a previous study [31].

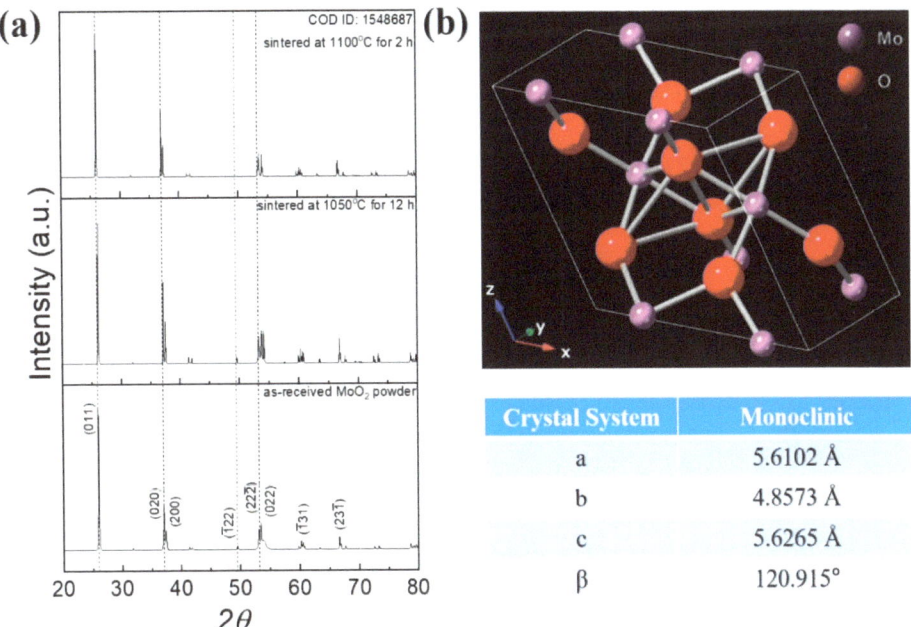

Figure 3. (**a**) XRD patterns and (**b**) crystal structure modeling and lattice parameters of the as-received and sintered samples.

The microstructures of the fracture cross-section of the sintered samples at different times and histograms of the grain size distribution are shown in Figure 4. By increasing the sintering time at 1050 °C, the as-received MoO$_2$ particles transformed from an irregular polygonal structure (Figure 1a) to those with a regular geometry with a nearly spherical structure (Figure 4c,d). After elapsing the sintering time of 2 h at 1050 °C, roughly rounded grains were roughly found in small and large grains, suggesting that a concave surface acts to pull itself into a flat surface to decrease the surface energy. Subsequently, the presence of surface tension forces induced by the difference in the curvature of the powder, which tends to minimize the surface area of materials, can cause the material to adopt a more rounded shape. The histograms of the grain size distribution of the sintered samples at different times show that the grains grow with increasing sintering time, and finally, the mean grain size reached 4.7 µm and the standard deviation of the grain sizes slightly increased. Grain growth refers to a particle's volume change by grain boundary motion and its driving force is proportional to the mean of curvature on the grain boundary. The convex grain boundaries of smaller particles move inward, and smaller particles shrink. Therefore, as the sintering time increases, the standard deviation of the grain sizes increases because the radius of the curved grain boundary reduces when smaller particle shrinks [32].

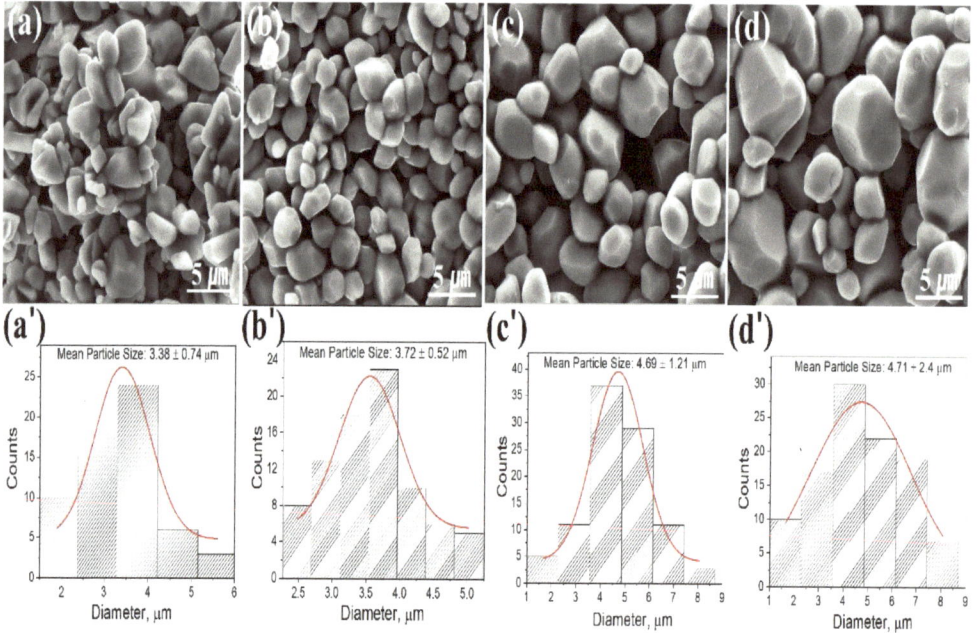

Figure 4. SEM photographs of the fracture cross-section of the sintered MoO$_2$ sample at 1050 °C for (**a**) 0.5, (**b**) 2, (**c**) 6, (**d**) 12 h; (**a′–d′**) show the histograms of grain size distribution.

Figure 5 shows the microstructures and grain size variation in the sintered samples' fracture cross-sections with increasing sintering temperatures for 2 h. Similar to Figure 4, the as-received MoO$_2$ powder with an irregular polygonal structure exhibits a partially further spherical shape when the sintering temperature was increased above 1000 °C, owing to the generation of surface tension force, capillary action, and particle rearrangement, as previously mentioned. Unlike Figure 4, particles with a partially spherical shape were observed in Figure 5a at a lower sintering temperature of 1000 °C for 2 h, indicating that the transformation from the irregular polygonal structure to a rounded shape is caused by sintering for sufficient time as well as over the temperature. By increasing the sintering temperature, the grain size distribution increased, as shown in Figure 5a′–d′. Higher sintering temperatures provide more thermal energy for grain boundary migration and coalescence, resulting in larger grain sizes and an increase in grain size distribution.

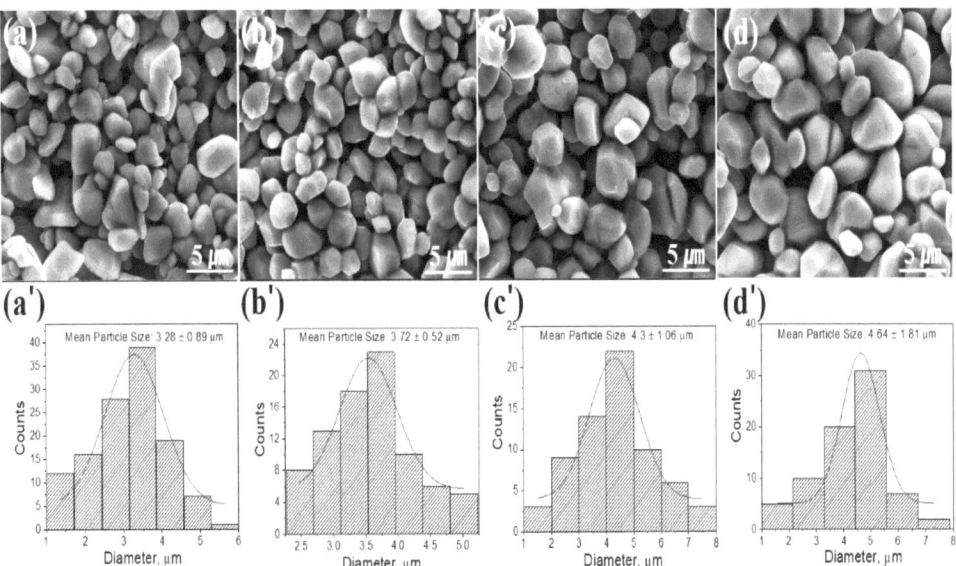

Figure 5. SEM photographs of fracture cross-section of sintered MoO$_2$ sample depending on different temperatures of (**a**) 1000, (**b**) 1050, (**c**) 1100, (**d**) 1150 °C for 2 h; (**a'**–**d'**) show the histograms of grain size distribution.

Based on the XRD patterns (Figure 2a), the crystallite size was calculated using the Scherrer equation, which considers the broadening of a peak in a diffraction pattern to relate the size of sub-micrometer crystallites as follows [33,34]:

$$D_{hkl} = \frac{C\lambda}{B_{hkl} \cdot \cos\theta}, \delta = 1/D_{hkl}^2 \qquad (1)$$

where D_{hkl} is the crystallite size in the direction perpendicular to the lattice planes, C is a numerical factor frequently referred to as the crystallite-shape factor and $C = 0.9$ is a good approximation [35], λ is the wavelength of the X-rays, B_{hkl} is the width (full-width at half-maximum) of the XRD peak in radians, θ is the Bragg angle, and δ is the dislocation density. Figure 6a,b show the variation in mean crystallite sizes and dislocation density, obtained from the Scherrer equation, according to the annealing time and temperature, respectively. Based on the obtained graph, the mean crystallite size increased, and dislocation density decreased with the sintering time and temperature. As shown in Figure 6a, crystallite growth was rapid in the initial stage and gradual after the intermediate stage. During the initial stages of heating, the crystallite growth rate increases rapidly, and concurrently the dislocation density decreases swiftly. However, as the crystallites increase in size, they begin to hinder crystallite growth owing to the decrease in dislocation density; thus, the growth rate decreased. Eventually, the growth rate decreased until the decrease rate in dislocation density was effectively negligible. As shown in Figure 6b, the growth of the mean crystallite size depends on the sintering temperature, and it was correlated with the variation in the dislocation density. At the same sintering time of 2 h, the mean crystallite size increased linearly with the decrease in dislocation density, indicating that the crystallite growth is correlated with the variation in the dislocation density. Dos Reis et al. [36] also show that the dislocation plays as an effective fast diffusivity channel in MgO based on kinetic Monte Carlo simulations testified by atomistic calculations.

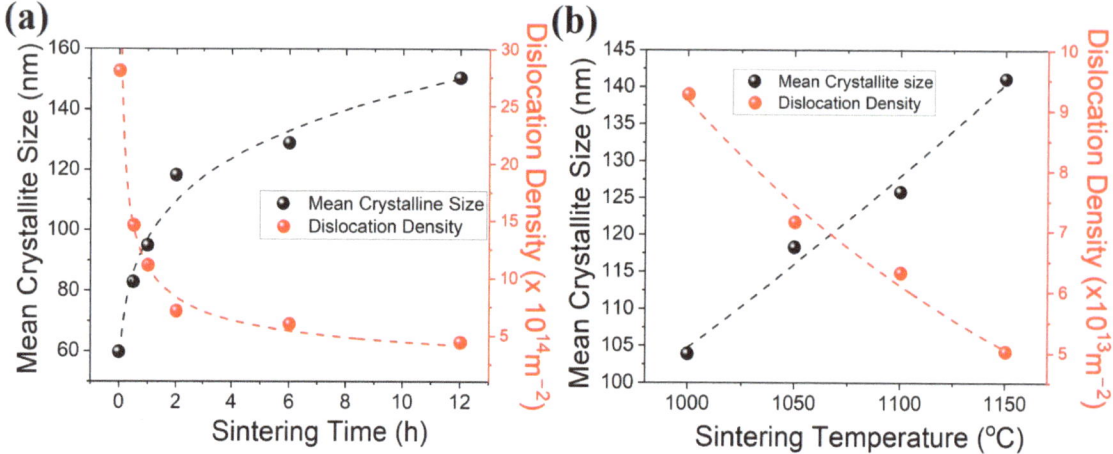

Figure 6. Evolution of the mean crystallite size and dislocation density of the samples as a function of (**a**) various holding times at a sintering temperature of 1050 °C and (**b**) various sintering temperatures for 2 h.

Figure 7 shows the variation in the average grain size of sintered samples with various sintering times at 1050 °C and various sintering temperatures for 2 h. As shown in Figure 7a, the average grain size sharply increased for the sintering time of 2 h, and then steadily increased until 12 h at the same sintering temperature. The graph shows that the grain growth rate is very high in the first 0.5 to 2 h, and then the growth rate decreases. Grain growth processes are also related to the mechanisms of grain boundary motion; thus, it is assumed that the instantaneous rate of growth is directly proportional to the instantaneous average rate of grain boundary migration in the structure, calculated as [37]

$$\frac{d\overline{G}}{dt} \sim \overline{v} \tag{2}$$

where $\overline{G}$ is the average grain size and $\overline{v}$ is the velocity of grain boundary motion. A grain boundary proceeds with a velocity ($\overline{v}$) in response to the net pressure on the boundary. It is generally given that the velocity is directly proportional to the pressure, with the constant of proportionality being the grain boundary mobility (M), and thus,

$$\overline{v} = MP \tag{3}$$

The driving force for boundary movement in ceramics is mostly acquired from the pressure gradient ΔP across the boundary arising from its curvature, given by [38]

$$\Delta P = S\left(\frac{1}{r_1} + \frac{1}{r_2}\right) = S\left(\frac{1}{K\overline{G}}\right) \tag{4}$$

where S is the grain boundary energy, r_1, and r_2 are the two radii of curvature of the boundary surface, and K is a constant. The pressure gradient ΔP decreases as the mean grain size $\overline{G}$ increases as a function of sintering time. Consequently, the reduction in ΔP could also contribute to the decrease in the velocity of grain boundary motion at the intermediate and final steps of the sintering process. By increasing the sintering temperature, grain coarsening occurs because higher temperatures provide more thermal energy for grain boundary movement and coalescence, resulting in larger grain sizes.

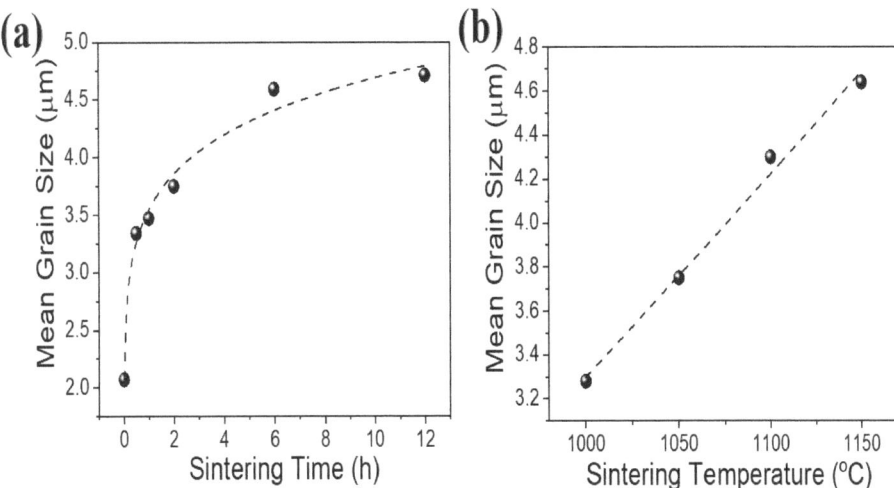

Figure 7. Changes in the mean grain size of the samples as a function of (**a**) various holding times at a sintering temperature of 1050 °C and (**b**) various sintering temperatures for 2 h.

Using the data on the crystallite and grain size variation with the sintering processes (Figures 6 and 7), the kinetic and activation energy of grain growth can be evaluated as shown in Figure 7. The grain (or crystallite) growth can be calculated as follows [39].

$$G - G_0 = K't^{1/n} \quad (5)$$

where G is the mean grain size, G_0 is the initial grain size, K' is a temperature-dependent constant, t is the time, and n is the grain growth exponent representing the grain growth behavior. Here, D on the left of the Y-axis of Figure 7a is the mean crystallite size. The n value can be calculated using the data analysis software of Origin (Origin Lab, Northampton, MA, USA) using an allometric equation (y = ax^b). Figure 7a shows that the m values for crystallite and grain growth are approximately 2.8 and 4, respectively. According to the literature [38], the values of the grain growth exponent n have been determined in the kinetics of grain growth for various mechanisms: n values of lattice and surface diffusion for pore-controlled systems are 3 and 4, respectively, and $n = 2$ for the boundary-controlled system. In this study, it is reasonable to assume that the n value of 2.8 for crystallite growth is mainly controlled by the lattice diffusion, and that of 4 for grain growth is predominantly governed by surface diffusion for pore-controlled systems. The activation energy (Q) of grain growth can be obtained by [39]

$$K' = K'_0 \exp\left(-\frac{Q}{RT}\right) \quad (6)$$

where Q is the activation energy, T is the temperature in Kelvin, K'_0 is the pre-exponential rate constant, and R is the gas constant. This explains why a plot of $ln(G - G_0)$ vs. $1/T$ produces a straight line (Figure 8b). The crystallite and grain growth activation energies are 95.62 and 76.96 kJ/mol, respectively. The grain growth representing the exponent n with a value of approximately 4 is the system where pore mobility is controlled by surface diffusion [38]. The activation energy of the grain growth should indicate the activation energy of the pore mobility controlled by surface diffusion in the sintered MoO_2 samples. On the contrary, an exponent of $n \approx 3$ describes the activation energy governed by lattice diffusion. Based on the interpretation, the value of the calculated activation energy of the crystallite and grain growth could be accepted given that crystallite growth is controlled by lattice diffusion through a dislocation channel, and grain growth is governed by surface diffusion of a pore mobility. Then, the value of activation energy for crystallite growth is greater than that for grain growth. However, it is difficult to identify a grain growth mechanism based on the exponent n alone, because Ganapathi et al. [29] obtained excellent fits for all values of n = 2, 3, or 4 with different annealing temperatures and times based on measurements on nanocrystalline Cu. Nevertheless, it is notable that the comparison between the kinetics and activation energies for the crystallite and grain growth could reliably be evaluated by the large reduction of some intrinsic limitations because of the same experimental conditions, namely, model parameters, acquisition temperature and time, purity, and usage of the same materials.

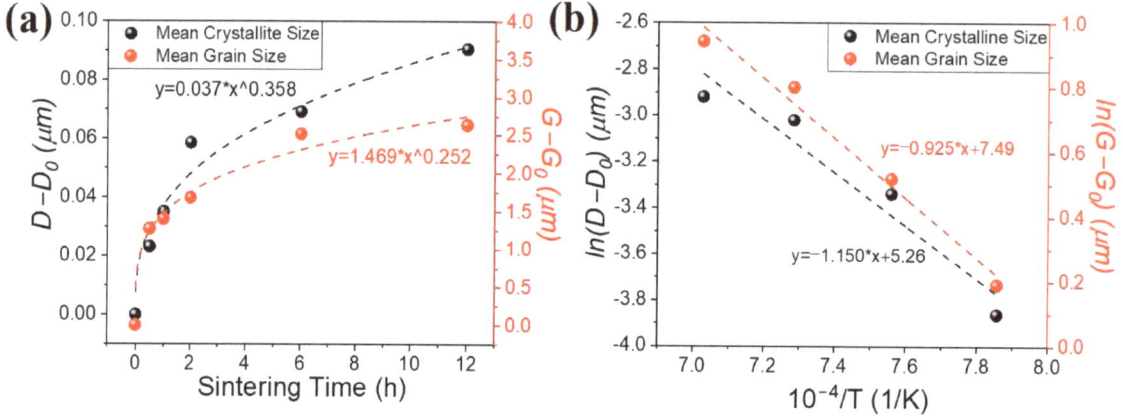

Figure 8. Variation of (**a**) $D - D_0$ and $G - G_0$ with different sintering times at 1050 °C and (**b**) plots of $ln(D - D_0)$ and $ln(G - G_0)$ vs. $1/T$ for different sintering temperatures at the durations of 2 h.

4. Conclusions

Compacted MoO_2 powders by dray pressing were sintered at various temperatures and times in an N_2 atmosphere, and their microstructure evolution was characterized using XRD and SEM images. As the sintering process reached a certain temperature and time, the irregular shape of as-received MoO_2 powders transformed into an equiaxed structure, owing to the occurrence of surface tension force, following which, the grain started to grow to decrease the surface energy. In accord with the Scherrer equation based on the XRD pattern, the crystallite size increased gradually as the sintering time and temperature increased, and the variation in grain size with sintering processes exhibited analogous trends as the change in the crystallite size. The crystallite and grain size exponentially increased with a growth exponent, n, of approximately 2.8 and 4, respectively, indicating that the growth kinetic can be governed by dislocation-mediated lattice diffusion and surface diffusion-controlled pore mobility, respectively. With increasing sintering temperature, both crystallite and grain size obeyed the Arrhenius equation against $1/T$, and the activation energies were 95.65 and 76.95 kJmol^{-1}, respectively.

Author Contributions: Conceptualization, J.L. and J.J. (Jinyoung Jeong); methodology, J.J. (Jinyoung Jeong), H.J. and J.J. (Jinman Jang); software, J.L.; validation, J.L.; formal analysis, J.J. (Jinyoung Jeong); investigation, J.J. (Jinyoung Jeong) and J.J. (Jinman Jang); resources, H.L. and J.P.; data curation, J.J. (Jinman Jang); writing—review and editing, J.L.; supervision, H.L. and J.P.; project administration, H.L. and J.P.; funding acquisition, J.P. All authors have read and agreed to the published version of the manuscript.

Funding: This research was funded by the Ministry of Trade, Industry and Energy, grant number 20022463, Korea.

Data Availability Statement: Not applicable.

Conflicts of Interest: The authors declare no conflict of interest.

References

1. De Castro, I.; Datta, R.; Ou, J.; Castellanos-Gomez, A.; Sriram, S.; Daeneke, T.; Kalantar-Zadeh, K. Molybdenum Oxides—From Fundamentals to Functionality. *Adv. Mater.* **2017**, *29*, 1701619. [CrossRef]
2. Yin, H.; Kuwahara, Y.; Mori, K.; Cheng, H.; Wen, M.; Yamashita, H. High-surface-area plasmonic MoO3−x: Rational synthesis and enhanced ammonia borane dehydrogenation activity. *J. Mater. Chem. A* **2017**, *5*, 8946–8953. [CrossRef]
3. Xia, W.; Xu, F.; Zhu, C.; Xin, H.; Xu, Q.; Sun, P.; Sun, L. Sea urchin-like NiCoO2@C nanocomposites for Li-ion batteries and supercapacitors. *Nano Energy* **2016**, *27*, 457–465. [CrossRef]
4. Bessonov, A.; Kirikova, M.; Petukhov, D.; Allen, M.; Ryhänen, T.; Bailey, M. Layered memristive and memcapacitive switches for printable electronics. *Nat. Mater.* **2015**, *14*, 199–204. [CrossRef] [PubMed]
5. Bin, X.; Tian, Y.; Luo, Y.; Sheng, M.; Luo, Y.; Ju, M.; Que, W. High-performance flexible and free-standing N-doped Ti3C2Tx/MoOx films as electrodes for supercapacitors. *Electrochim. Acta* **2021**, *389*, 138774. [CrossRef]
6. Brewer, L.; Lamoreaux, R. The Mo-O system (Molybdenum-Oxygen). *Bull. Alloy Phase Diagr.* **1980**, *1*, 85–89. [CrossRef]
7. Clentsmith, G.; Cloke, F.; Green, J.; Hanks, J.; Hitchcock, P.; Nixon, J. Stabilization of Low-Oxidation-State Early Transition-Metal Complexes Bearing 1,2,4-Triphosphacyclopentadienyl Ligands: Structure of [{Sc(P3C2tBu2)2}2]; ScII or Mixed Oxidation State? *Angew. Chem. Int. Ed. Engl.* **2003**, *42*, 1068–1071. [CrossRef]
8. Scanlon, D.; Watson, G.; Payne, D.; Atkinson, G.; Egdell, R.; Law, D. Theoretical and Experimental Study of the Electronic Structures of MoO3 and MoO2. *J. Phys. Chem. C* **2010**, *114*, 4636–4645. [CrossRef]
9. Magnéli, A.; Andersson, G. Studies on the Hexagonal Tungsten Bronzes of Potassium, Rubidium, and Cesium. *Acta Chem. Scand.* **1995**, *9*, 315–324. [CrossRef]
10. Zhao, X.; Cao, M.; Liu, B.; Tian, Y.; Hu, C. Interconnected core–shell MoO2 microcapsules with nanorod-assembled shells as high-performance lithium-ion battery anodes. *J. Mater. Chem.* **2012**, *22*, 13334–13340. [CrossRef]
11. Shi, Y.; Guo, B.; Corr, S.; Shi, Q.; Hu, Y.; Heier, K.; Chen, L.; Seshadri, R.; Stucky, G. Ordered Mesoporous Metallic MoO2 Materials with Highly Reversible Lithium Storage Capacity. *Nano Lett.* **2009**, *9*, 4215–4220. [CrossRef]
12. Miyata, N.; Akiyoshi, S. Preparation and electrochromic properties of rf-sputtered molybdenum oxide films. *J. Appl. Phys.* **1985**, *58*, 1651–1655. [CrossRef]
13. Mohamed, S.; Kappertz, O.; Ngaruiya, J.; Leervad Pedersen, T.; Drese, R.; Wuttig, M. Correlation between structure, stress and optical properties in direct current sputtered molybdenum oxide films. *Thin Solid Films* **2003**, *429*, 135–143. [CrossRef]
14. Zhang, W.; Desikan, A.; Oyama, S. Effect of Support in Ethanol Oxidation on Molybdenum Oxide. *J. Phys. Chem.* **1995**, *99*, 14468–14476. [CrossRef]
15. Katrib, A.; Leflaive, P.; Hilaire, L.; Maire, G. Molybdenum based catalysts. I. MoO2 as the active species in the reforming of hydrocarbons. *Catal. Lett.* **1995**, *38*, 95–99. [CrossRef]
16. Wang, Z.; Zhang, C.; Chen, D.; Tang, S.; Zhang, J.; Wang, Y.; Han, G.; Xu, S.; Hao, Y. Flexible ITO-Free Organic Solar Cells Based on MoO3/Ag Anodes. *IEEE Photonics J.* **2015**, *7*, 8400109. Available online: https://ieeexplore.ieee.org/document/7021946 (accessed on 1 August 2023).
17. Wang, S.; Guo, H.; Song, Y.; Ma, G.; Li, S.; Zhang, L.; Shao, X. P-63: Lower Reflective TFT Materials and Technology Innovation. *SID Symp. Dig. Tech. Pap.* **2017**, *48*, 1478–1481. [CrossRef]
18. Fernandes Cauduro, A.L.; Fabrim, Z.E.; Ahmadpour, M.; Fichtner, P.F.P.; Hassing, S.; Rubahn, H.-G.; Madsen, M. Tuning the optoelectronic properties of amorphous MoOx films by reactive sputtering. *Appl. Phys. Lett.* **2015**, *106*, 202101. [CrossRef]
19. Liu, Y.; Zhang, H.; Ouyang, P.; Chen, W.; Wang, Y.; Li, Z. High electrochemical performance and phase evolution of magnetron sputtered MoO2 thin films with hierarchical structure for Li-ion battery electrodes. *J. Mater. Chem. A* **2014**, *2*, 4714–4721. [CrossRef]
20. Qiu, L.; Chen, K.; Yang, D.; Zhang, M.; Hao, X.; Li, W.; Zhang, J.; Wang, W. Metal copper induced the phase transition of MoO3 to MoO2 thin films for the CdTe solar cells. *Mater. Sci. Semicond.* **2021**, *122*, 105475. [CrossRef]
21. Fujiwara, K.; Tsukazaki, A. Formation of distorted rutile-type NbO2, MoO2, and WO2 films by reactive sputtering. *J. Appl. Phys.* **2019**, *125*, 085301. [CrossRef]
22. Ahn, E.; Lee, J.; Koh, Y.; Lee, J.; Park, B.; Kim, J.; Lee, I.; Lee, C.; Jeen, H. Low Temperature Nanoscale Oxygen-Ion Intercalation into Epitaxial MoO2 Thin Films. *J. Phys. Chem. C* **2017**, *121*, 3410–3415. [CrossRef]
23. Martínez, M.A.; Guillén, C. Comparison between large area dc-magnetron sputtered and e-beam evaporated molybdenum as thin film electrical contacts. *J. Mater. Process. Technol.* **2003**, *143–144*, 326–331. [CrossRef]
24. Pachlhofer, J.; Martín-Luengo, A.; Franz, R.; Franzke, E.; Köstenbauer, H.; Winkler, J.; Bonanni, A.; Mitterer, C. Industrial-scale sputter deposition of molybdenum oxide thin films: Microstructure evolution and properties. *J. Vac. Sci. Technol. A* **2017**, *35*, 021504. [CrossRef]
25. Park, H.; Ryu, J.; Youn, H.; Yang, J.; Oh, I. Fabrication and Property Evaluation of Mo Compacts for Sputtering Target Application by Spark Plasma Sintering Process. *Mater. Trans.* **2012**, *53*, 1056–1061. [CrossRef]
26. Durnez, A.; Petitbon-Thévenet, W.; Fortuna, F.; Radioanal, J. Preparation of molybdenum target by centrifugal method. *J. Radioanal. Nucl. Chem.* **2014**, *299*, 1149–1154. [CrossRef]
27. Wang, Y.; Tang, Q.; Chen, D.; Liu, X.; Xiong, X. Microstructure and Magnetron Sputtering Properties of Molybdenum Target Prepared by Low-Pressure Plasma Spraying. *J. Therm Spray Technol.* **2019**, *28*, 1983–1994. [CrossRef]
28. Kitchamsetti, N.; Choudhary, R.; Phase, D.; Devan, R. Structural correlation of a nanoparticle-embedded mesoporous CoTiO3 perovskite for an efficient electrochemical supercapacitor. *RSC Adv.* **2020**, *10*, 23446. [CrossRef]

29. Ganapathi, S.; Owen, D.; Chokshi, A. The kinetics of grain growth in nanocrystalline copper. *Scr. Metall. Mater.* **1991**, *25*, 2699–2704. [CrossRef]
30. Adrian, A.B.; Brendan, J.K.; Christopher, J.H. Neutron Powder Diffraction Study of Molybdenum and Tungsten Dioxides. *Aust. J. Chem.* **1995**, *48*, 1473–1477. [CrossRef]
31. Jacob, K.; Saji, V.; Gopalakrishnan, J.; Waseda, Y. Thermodynamic evidence for phase transition in MoO_{2-d}. *J. Chem. Thermodyn.* **2007**, *39*, 1539–1545. [CrossRef]
32. Wakai, F.; Yoshida, M.; Shinoda, Y.; Akatsu, T. Coarsening and grain growth in sintering of two particles of different sizes. *Acta Metall.* **2005**, *53*, 1361–1371. [CrossRef]
33. Scherrer, P. Bestimmung der Grosse und inneren Struktur von Kolloidteilchen mittels Rontgenstrahlen. *Nach. Ges. Wiss. Gottingen.* **1918**, *2*, 98–100.
34. Holzwarth, U.; Gibson, N. The Scherrer equation versus the 'Debye-Scherrer equation'. *Nat. Nanotechnol.* **2011**, *6*, 534. [CrossRef] [PubMed]
35. Klug, H.; Alexander, L. *X-ray Diffraction Procedures: For Polycrystalline and Amorphous Materials*, 2nd ed.; Wiley: Hoboken, NJ, USA, 1974. Available online: https://ui.adsabs.harvard.edu/abs/1974xdpf.book.....K/abstract (accessed on 1 August 2023).
36. Dos Reis, M.; Giret, Y.; Carrez, P.; Cordier, P. Efficiency of the vacancy pipe diffusion along an edge dislocation in MgO. *Comput. Mater. Sci.* **2022**, *211*, 111490. [CrossRef]
37. Burke, J.; Turnbull, D. Recrystallization and grain growth. *Prog. Phys. Met.* **1952**, *3*, 275–292. [CrossRef]
38. Wang, F. *Ceramic Fabrication Processes: Treatise on Materials Science and Technology*; Academic Press Inc.: Orlando, FL, USA, 1976; Volume 9. Available online: https://books.google.co.kr/books?hl=en&lr=&id=00MvBQAAQBAJ&oi=fnd&pg=PP1&ots=-hofU5QfJ6&sig=MtR1oCG1XwI-BY5xhn7jhpW39ZA&redir_esc=y#v=onepage&q&f=false (accessed on 1 August 2023).
39. Lu, K. Nanocrystalline metals crystallized from amorphous solids: Nanocrystallization, structure, and properties. *Mater. Sci. Eng. R.* **1996**, *16*, 161–221. [CrossRef]

Disclaimer/Publisher's Note: The statements, opinions and data contained in all publications are solely those of the individual author(s) and contributor(s) and not of MDPI and/or the editor(s). MDPI and/or the editor(s) disclaim responsibility for any injury to people or property resulting from any ideas, methods, instructions or products referred to in the content.

Article

Properties of Perovskite-like Lanthanum Strontium Ferrite Ceramics with Variation in Lanthanum Concentration

Daryn B. Borgekov [1,2], Artem L. Kozlovskiy [1,2,*], Rafael I. Shakirzyanov [2], Ainash T. Zhumazhanova [1,2], Maxim V. Zdorovets [1,2] and Dmitriy I. Shlimas [1,2]

[1] Laboratory of Solid State Physics, The Institute of Nuclear Physics, Almaty 050032, Kazakhstan
[2] Engineering Profile Laboratory, L.N. Gumilyov Eurasian National University, Nur-Sultan 010008, Kazakhstan
* Correspondence: kozlovskiy.a@inp.kz; Tel./Fax: +7-7024413368

Abstract: The purpose of this work is to study the effect of lanthanum (La) concentration on the phase formation, conductivity, and thermophysical properties of perovskite-like strontium ferrite ceramics. At the same time, the key difference from similar studies is the study of the possibility of obtaining two-phase composite ceramics, the presence of various phases in which will lead to a change in the structural, strength, and conductive properties. To obtain two-phase composite ceramics by mechanochemical solid-phase synthesis, the method of the component molar ratio variation was used, which, when mixed, makes it possible to obtain a different ratio of elements and, as a result, to vary the phase composition of the ceramics. Scanning electron microscopy, X-ray phase analysis, and impedance spectroscopy were used as research methods, the combination of which made it possible to comprehensively study the properties of the synthesized ceramics. Analysis of phase changes depending on lanthanum concentration change can be written as follows: $(La_{0.3}Sr_{0.7})_2FeO_4/LaSr_2Fe_3O_8 \rightarrow (La_{0.3}Sr_{0.7})_2FeO_4/LaSr_2Fe_3O_8/Sr_2Fe_2O_5 \rightarrow (La_{0.3}Sr_{0.7})_2FeO_4/Sr_2Fe_2O_5$. Results of impedance spectroscopy showed that with an increase in lanthanum concentration from 0.10 to 0.25 mol in the synthesized ceramics, the value of the dielectric permittivity increases significantly from 40.72 to 231.69, the dielectric loss tangent increases from 1.07 to 1.29 at a frequency of 10,000 Hz, and electrical resistivity decreases from 1.29×10^8 to 2.37×10^7 $\Omega \cdot$cm.

Keywords: perovskites; strontium lanthanum ferrite; phase transformations; hardening; solid-phase synthesis

1. Introduction

One of the promising areas of research in the field of alternative energy is the search for new types of solid oxide fuel cells (SOFC) capable of operating at lower temperatures than classical materials. Interest in this area is due to the possibility of reducing the operating temperatures of SOFC materials, as well as great prospects for reducing the cost and energy costs [1–3]. At the same time, reducing the operating temperatures of SOFC elements will not only reduce energy consumption and increase efficiency, but first of all, increase the service life of the material [4,5].

In this regard, in the past few years, much attention has been paid to the development and testing of various hypotheses in this area of research, most of which are related to the study of the physicochemical, thermophysical, and conductive properties of various materials—candidates for SOFC [6,7]. At the same time, despite serious shortcomings in performance during operation at low temperatures, interest in these types of solid fuel cells does not weaken due to the large potential for reducing energy costs in the case of solving several issues related to the reduction of oxygen and hydrogen from cathode materials at low temperatures. One of the promising materials in this direction is perovskite or perovskite-like structures, which will serve as an alternative to traditional SOFC elements based on compounds of lanthanum strontium manganate with yttrium-stabilized zirconium dioxide. Among such materials, perovskite ceramics based on lanthanum strontium ferrite are distinguished, the interest in which is due to their mixed type of electron-ion conductivity, as well as the possibility of effective oxygen reduction when used as cathode materials [8–11]. Interest in them is also due to the fact that the use of this class of ferrites can help reduce SOFC operating temperatures from high (800–1000 °C) to intermediate and low temperatures (400–800 °C). At the same time, a decrease in operating temperatures will make it possible to achieve an increase in the durability and stability of the SOFC material to degradation, as well as an improvement in thermal compatibility and stability of performance for a long time [11–15].

However, as is known, lanthanum strontium ferrite compounds have a wide range of composition stoichiometry variations, with a change in which there are significant changes not only in structural, but also in conducting and stoichiometric properties [16,17]. In this connection, studies in this direction, related to the study of the influence of stoichiometry, as well as changes in the phase composition under various synthesis conditions, have been very relevant in recent times, since the results of such studies will not only provide new data on the properties of these perovskite compounds, but also answer a number of questions related to both the phase composition effect on the electrochemical and conductive properties, and the resistance of ceramics to external influences. At the same time, variation in stoichiometry and, consequently, phase composition can be due to both synthesis processes and changes in their conditions, and variations in the initial components during the creation of perovskites [18–20].

Based on the foregoing, the main aim of this work is to consider the effect of lanthanum concentration (La) change during the synthesis of perovskite-like ceramics based on strontium ferrite on phase formation and subsequent changes in the conductive and thermophysical parameters.

2. Experimental Part

The synthesis of perovskite-like ceramics based on lanthanum-doped strontium ferrite compounds was carried out using a mechanochemical solid-phase synthesis method.

Mechanochemical synthesis was carried out using a variation in the molar ratio of the initial $SrCO_3$, $FeSO_4 \times 7H_2O$, $La(NO_3)_3$ components according to the chosen scheme: $(0.5 - x/2)\ SrCO_3 : (0.5 - x/2)\ FeSO_4 \times 7H_2O : x\ La(NO_3)_3$, where x of the $La(NO_3)_3$ component ranged from 0.10 to 0.25 mol. After mechanochemical grinding, the resulting mixtures were subjected to thermal annealing at a temperature of 1000 °C. According to the classical chemical reactions of thermal decomposition of the initial components used to obtain perovskite ceramics, when these components are heated, gaseous compounds are released in the form of CO_2, SO_2, NO_2, and H_2O. The isolation of these compounds with the formation of simple Fe_2O_3, La_2O_3, SrO oxides, as a rule, occurs at temperatures above 700 °C. In this case, the formation of complex perovskites occurs through the interaction of Fe_2O_3, La_2O_3, SrO oxides.

Mechanochemical solid-phase synthesis was carried out using the PULVERISETTE 6 classic line planetary mill (Fritsch, Idar-Oberstein, Germany). Grinding was carried out at a speed of 400 rpm for 1 h. After grinding, the resulting mixtures were annealed at a temperature of 1000 °C for 5 h; the samples were cooled to room temperature for 24 h while they were inside the furnace.

The study of morphological features and changes in the shape and size of the grains of the synthesized ceramics depending on the La concentration was carried out using the scanning electron microscopy method implemented on a Jeol 7500 F microscope (Jeol, Tokyo, Japan). The element distribution isotropy in the structure of ceramics was determined using the method of energy dispersive analysis.

The phase composition and structural parameters of the synthesized ceramics depending on the component concentration, as well as with dopant variation, were estimated using the X-ray diffraction method implemented on a D8 Advance ECO powder diffractometer (Bruker, Berlin, Germany). The phase composition was determined using the PDF-2(2016) database. Refinement of the parameters and determination of the degree of crystallinity were carried out in the DiffracEVA v.4.2 program code.

Measurement of current–voltage characteristics using the cyclic voltammetry method was carried out at room temperature (T = 25 °C) using a potentiostat-impedance analyzer PalmSens4+ (PalmSens BV, Houten, The Netherlands). Polished copper plates of the same area were used as electrodes (1 cm^2). Impedance spectroscopy of the obtained samples was performed on a HIOKI IM3533-01 LCR meter on compressed powders. Silver glue was used to create electrical contacts. To measure the cyclic current–voltage characteristics and impedance characteristics, the resulting powders were pressed into pellets. All measurements were performed in five parallels to determine the measurement error, standard deviations, and also to establish the repeatability of the results obtained not only on different samples, but also when re-measuring the obtained characteristics to determine the stability. Figure 1 shows an image of powders obtained after mechanochemical solid-phase synthesis and thermal annealing, as well as after pressing to measure the impedance characteristics. To measure the characteristics, the samples were pressed into pellets 10 mm in diameter and 0.5 mm thick. The obtained samples after pressing were subjected to thermal drying at a temperature of 60 °C; for 2 h to evaporate the polyvinyl alcohol used for pressing.

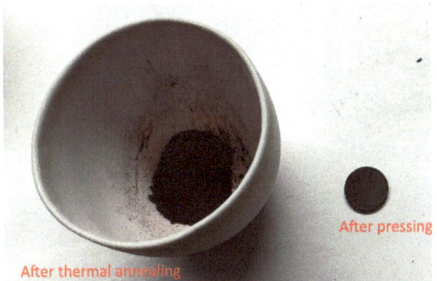

Figure 1. Image of perovskites before and after the pressing procedure.

3. Results and Discussion

One of the most common ways to obtain perovskite or perovskite-like ceramics is the method of mechanochemical solid-phase synthesis, whose condition variation makes it possible to obtain, with a high controllable accuracy, structures with specified parameters or phase composition in any volume, which can be used quite well in the future when scaling up this technology and manufacturing ceramics on an industrial scale. At the same time, the variation of the initial mixture composition also makes it possible to obtain different compositions of ceramics, with a uniform distribution of elements over the volume.

Figure 2 below shows SEM images of the studied perovskite-like ceramics depending on the lanthanum concentration, reflecting changes in the morphological features of the ceramics. The SEM images are made on the obtained powders after thermal annealing. As can be seen from the data presented, in the case of a lanthanum concentration of 0.10 mol, ceramics are coarse-grained structures, rhomboid or pyramidal in shape, surrounded by spherical or near-spherical particles, the size of which varies in the range of 150–300 nm. At the same time, analysis of the obtained images showed that quite large grains are also found in ceramics, the size of which exceeds 1–1.5 µm, which are surrounded by smaller spherical particles; in the aggregate, this structural phenomenon is close to a dendrite-like structure with high porosity. An increase in the lanthanum concentration to 0.15 mol in the structure of ceramics also leads to the formation of dendrite-like structures; however, with a slight decrease in the size of large particles, as well as the formation of double triple grain boundaries between particles.

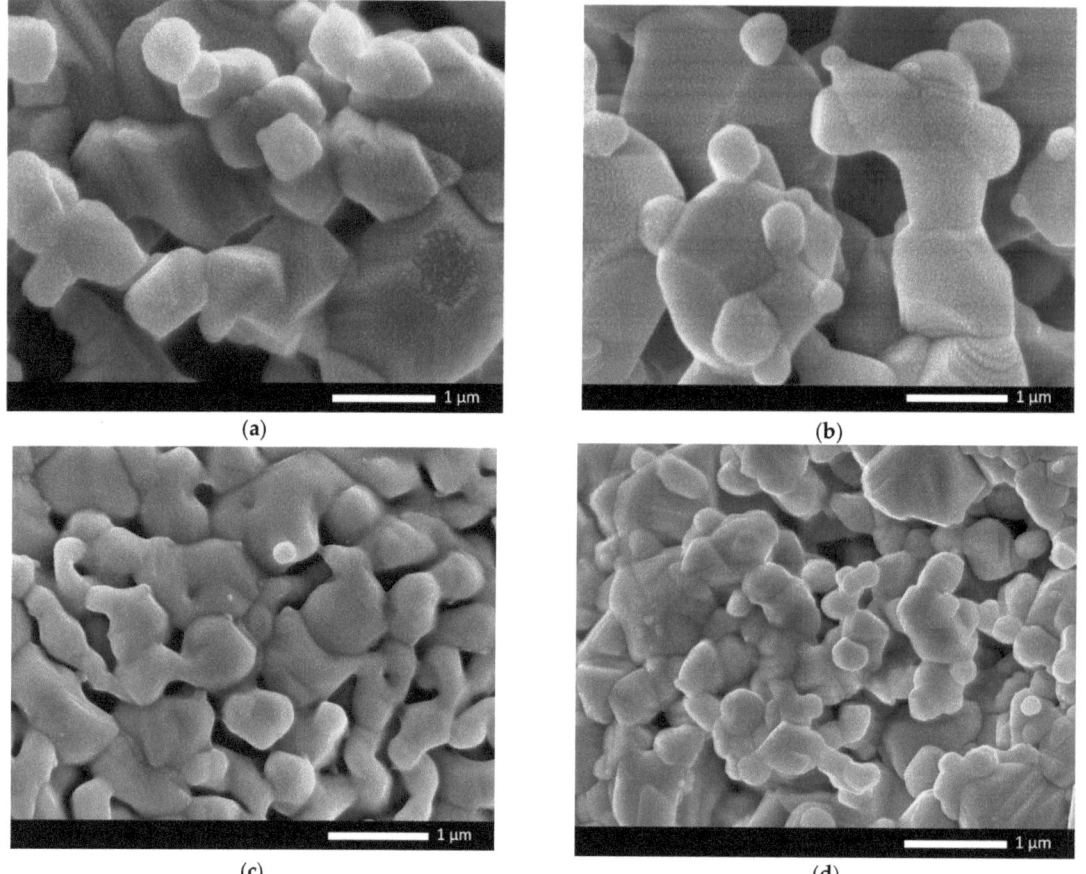

Figure 2. SEM images of synthesized perovskite-like powders as a function of La concentration: (**a**) 0.10 mol; (**b**) 0.15 mol; (**c**) 0.20 mol; (**d**) 0.25 mol.

With an increase in the lanthanum concentration to 0.20 mol and higher, a significant change in the morphological features of the synthesized perovskite-like ceramics is observed, which consists of the formation of a fine-grained structure that forms clusters of grains, as well as the absence of large grains, which were found at lower lanthanum concentrations. At the same time, the concentration of pores with a change in morphology and a decrease in grain size becomes much lower, and in the case of a concentration of 0.25 mol. It is also worth noting that a change in concentration leads to an increase in the degree of homogeneity of grain sizes, due to a decrease in the contribution of coarse-grained inclusions.

Table 1 presents the results of energy-dispersive analysis of the studied samples, reflecting the ratio of elements. As can be seen from the data presented, an increase in the lanthanum concentration for synthesis leads to an increase in its content in the composition of the samples under study.

A change in morphological features, as a rule, can be associated with a change in the phase composition of the samples as a result of the initialization of phase transformation processes with a change in the ratio of component concentrations in ceramics. The figure shows the results of X-ray phase analysis of the studied samples depending on the concentration of lanthanum, reflecting the processes of phase transformations with a change in the component concentration. The general view of the presented diffraction patterns, in addition to the change in the phase composition, which is reflected in the formation of new diffraction reflections with a change in the lanthanum concentration, also shows changes in the degree of crystallinity of the synthesized ceramics (see the data in Figure 3).

Table 1. Data of energy-dispersive analysis.

Concentration of Element, at. %	Concentration, mol			
	0.10	0.15	0.20	0.25
La	2.2 ± 0.3	4.6 ± 0.5	7.3 ± 0.5	10.4 ± 0.9
Sr	21.2 ± 1.6	20.1 ± 1.5	17.8 ± 1.1	16.4 ± 1.7
Fe	17.3 ± 1.2	16.6 ± 1.3	16.4 ± 1.8	15.3 ± 1.4
O	59.3 ± 2.2	58.7 ± 2.5	58.5 ± 2.6	57.9 ± 2.3

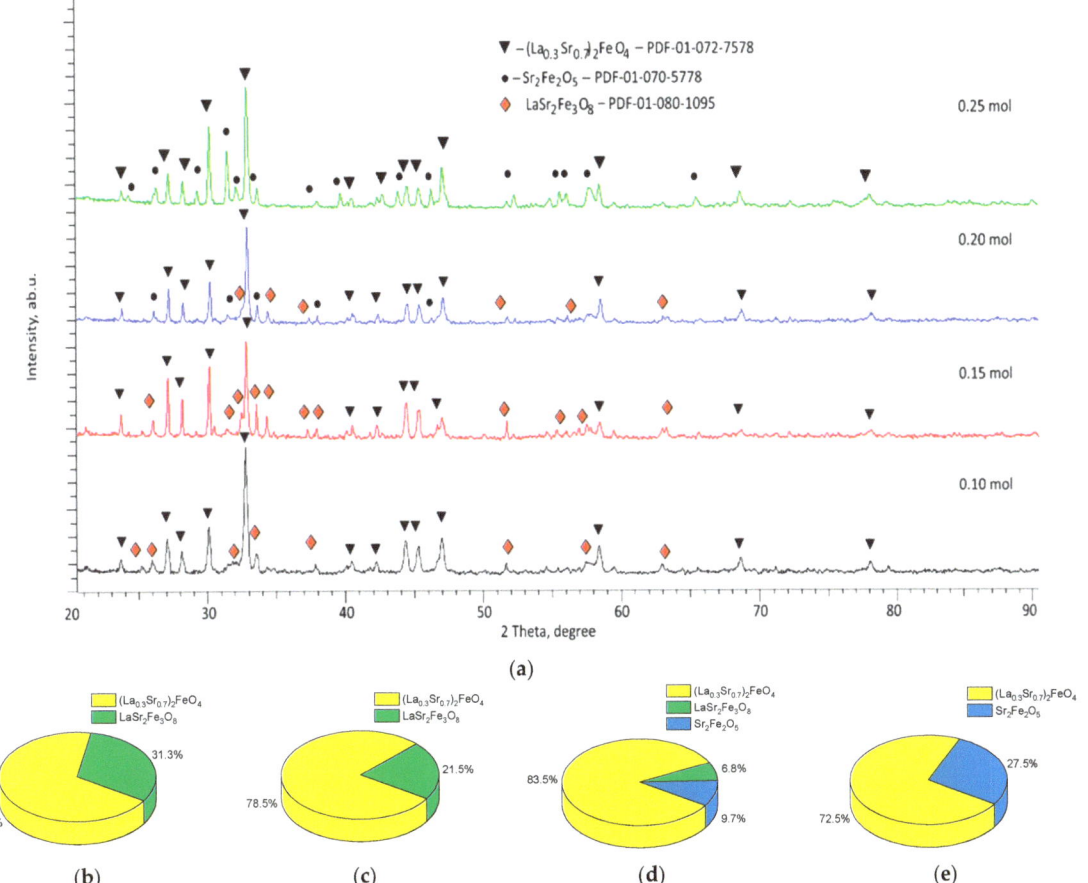

Figure 3. (a) X-ray diffraction patterns of synthesized ceramics based on lanthanum strontium ferrite depending on the lanthanum concentration; phase diagrams of the synthesized ceramics depending on the lanthanum concentration: (b) 0.10 mol; (c) 0.15 mol; (d) 0.20 mol; (e) 0.25 mol.

According to the phase composition analysis of the samples, it was found that at lanthanum concentrations of 0.10–0.15, the main contribution is made by the tetragonal $(La_{0.3}Sr_{0.7})_2FeO_4$ phase, as well as the orthorhombic $LaSr_2Fe_3O_8$ phase, the content of which decreases with an increase in lanthanum concentration from 31.3% to 21.5%. These phases are structures of substituted strontium ferrite, in which part of the strontium ions are replaced by lanthanum, in which partial substitution of Sr^{2+} for La^{3+} leads to the formation of structures with mixed iron valence, as well as the possible formation of oxygen vacancies. The contributions of each phase were estimated by determining the areas of reflections with subsequent calculation of their weight contribution in terms of 100%. At the same time, as can be seen from the analysis of the shape of diffraction reflections and their relationship with the area characteristic of an amorphous-like or disordered structure, a decrease in the $LaSr_2Fe_3O_8$ phase contribution leads to an increase in structural ordering (see the data in Figure 4).

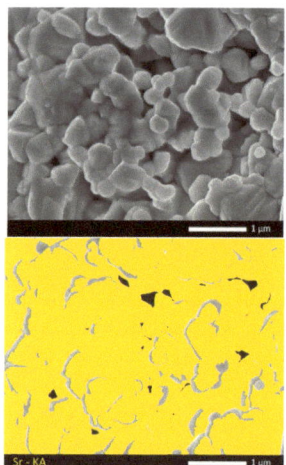
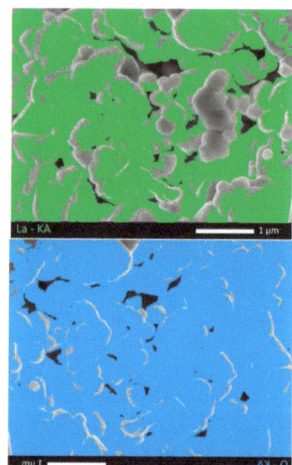
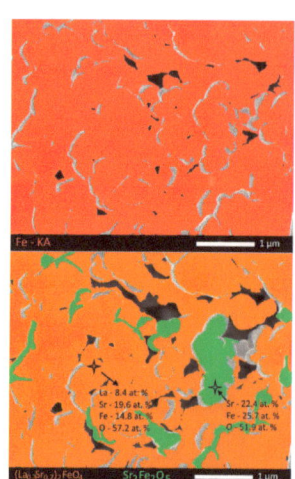

Figure 4. Mapping results of the studied powders with La concentration equal to 0.25 mol.

In the case of a change in concentration from 0.15 mol to 0.20 mol, in addition to the $LaSr_2Fe_3O_8$ phase displacement, which is expressed in a decrease in the intensity of diffraction reflections characteristic of it, the appearance of reflections characteristic of the orthorhombic $Sr_2Fe_2O_5$ phase (brownmillerite-type structure) was observed on the diffraction pattern, the presence of which indicates phase transformation processes occurring as a result of a change in the concentration of components, as well as the $(La_{0.3}Sr_{0.7})_2FeO_4$ phase dominance. At the same time, analysis of the shape of reflections indicates that these phase transformations are accompanied by an increase in the degree of crystallinity and a decrease in distortions and deformations in the structure of ceramics. In the case of a lanthanum concentration of 0.25 mol, the complete displacement of the $LaSr_2Fe_3O_8$ phase and the formation of two-phase ceramics, which has a high degree of crystallinity of more than 90% is observed. The presence of amorphous inclusions determined from the crystallinity degree values may be due to incompletely formed phases during thermal annealing, as well as transitional inclusions in the form of oxide compounds formed during the decomposition of the initial components.

Analyzing the obtained X-ray phase analysis data of the synthesized ceramics, depending on the lanthanum concentration, the following phase conversion scheme can be formed:

$(La_{0.3}Sr_{0.7})_2FeO_4/LaSr_2Fe_3O_8$ → $(La_{0.3}Sr_{0.7})_2FeO_4/LaSr_2Fe_3O_8/Sr_2Fe_2O_5$ → $(La_{0.3}Sr_{0.7})_2FeO_4/Sr_2Fe_2O_5$. The basic equations of chemical reactions that occur as a result of the thermal annealing of samples with the formation of those observed by X-ray diffraction can be written as follows (Equations (1)–(3)):

$$14SrCO_3 + 6La(NO_3)_3 + 5Fe_2(SO_4)_3 \rightarrow 10(La_{0.3}Sr_{0.7})_2FeO_4 + 14CO_2 + 18NO_2 + 15SO_2 + 13O_2 \quad (1)$$

$$4SrCO_3 + 2La(NO_3)_3 + 3Fe_2(SO_4)_3 \rightarrow 2LaSr_2Fe_3O_8 + 4CO_2 + 6NO_2 + 9SO_2 + 6O_2 \quad (2)$$

$$2SrCO_3 + Fe_2(SO_4)_3 \rightarrow Sr_2Fe_2O_5 + 2CO_2 + 3SO_2 + O_2 \qquad (3)$$

At the same time, Equations (1) and (2), according to the data of X-ray phase analysis, are typical for $La(NO_3)_3$ concentrations equal to 0.10–0.15 mol, and reactions Equations (1) and (3) for $La(NO_3)_3$ concentrations equal to 0.20–0.25 mol. The formation of the $Sr_2Fe_2O_5$ phase in the form of individual inclusions was also confirmed by the results of mapping the samples, according to which, in the structure of the annealed powders, the presence of small grains not containing lanthanum is observed, and the analysis of the elemental composition of these grains shows the ratio of Sr:Fe:O elements close to the stoichiometric ratio for the $Sr_2Fe_2O_5$ phase (see the data in Figure 4).

Table 2 presents the data on the change in the crystal lattice parameters depending on the La concentration. The density of ceramics was calculated using the method of X-ray diffraction analysis based on changing the structural parameters and volume of the crystal lattice, as well as taking into account the contribution of various phases, and for comparison, the results of measuring the density of ceramics obtained using the Archimedes method for pressed samples are given.

Table 2. Structural parameter data.

Phase	Concentration, mol			
	0.10	0.15	0.20	0.25
$(La_{0.3}Sr_{0.7})_2FeO_4$—Tetragonal I4/mmm(139)	a = 3.8561 ± 0.0026 Å, c = 12.6995 ± 0.0018 Å	a = 3.8661 ± 0.0029 Å, c = 12.7069 ± 0.0021 Å	a = 3.8548 ± 0.0024 Å, c = 12.6746 ± 0.0032 Å	a = 3.8495 ± 0.0025 Å, c = 12.6572 ± 0.0032 Å
$LaSr_2Fe_3O_8$—Orthorhombic Pmma(51)	a = 5.5214 ± 0.0027 Å, b = 11.9101 ± 0.0023 Å, c = 5.6127 ± 0.0024 Å	a = 5.5333 ± 0.0029 Å, b = 11.8751 ± 0.0023 Å, c = 5.6226 ± 0.0021 Å	a = 5.5279 ± 0.0024 Å, b = 11.8588 ± 0.0022 Å, c = 5.6127 ± 0.0024 Å	-
$Sr_2Fe_2O_5$—Orthorhombic, Icmm(74)	-	-	a = 5.6575 ± 0.0032 Å, b = 15.5625 ± 0.0023 Å, c = 5.5216 ± 0.0034 Å	a = 5.6519 ± 0.0023 Å, b = 15.5533 ± 0.0025 Å, c = 5.5118 ± 0.0034 Å
Density (method of X-ray diffraction analysis), g/cm³	5.673	5.694	5.654	5.535
Density (Archimedes method), g/cm³	5.564	5.523	5.602	5.504

According to the presented data, it can be seen that the density values obtained by both methods are in fairly good agreement, which indicates a rather low porosity in the case of pressed samples. In this case, the change in density has a pronounced dependence on the phase composition, and a slight decrease in density is due to the lower density for the $Sr_2Fe_2O_5$ phase in comparison with the density of $LaSr_2Fe_3O_8$.

An analysis of the change in the crystal lattice parameters depending on the lanthanum concentration indicates that the phase transformation processes are accompanied by the crystal lattice rearrangement, with partial substitution of ions at the sites, as well as their intrusion into interstices, which leads to deformation processes of stretching and volume expansion, which are clearly visible when the concentration increases from 0.10 to 0.15 mol. Additionally, an increase in the parameters for a given concentration range can be explained by the formation of an orthorhombic $Sr_2Fe_2O_5$ phase in the structure. Displacement of the $LaSr_2Fe_3O_8$ phase at concentrations of 0.20 mol and higher leads to a decrease in the crystal lattice parameters, which indicates structural ordering, which is also clearly visible when analyzing the shape of diffraction reflections and calculating the crystallinity degree parameter, the results of which change depending on the lanthanum concentration are shown in Figure 5. Additionally, during analysis of the deformation contributions to the change in structural parameters, as well as their distortion, it was found that the

$LaSr_2Fe_3O_8$ phase displacement leads to a more than threefold decrease in the defect fraction concentration in the ceramic structure, which can also contribute to a change in its conducting and electrochemical properties.

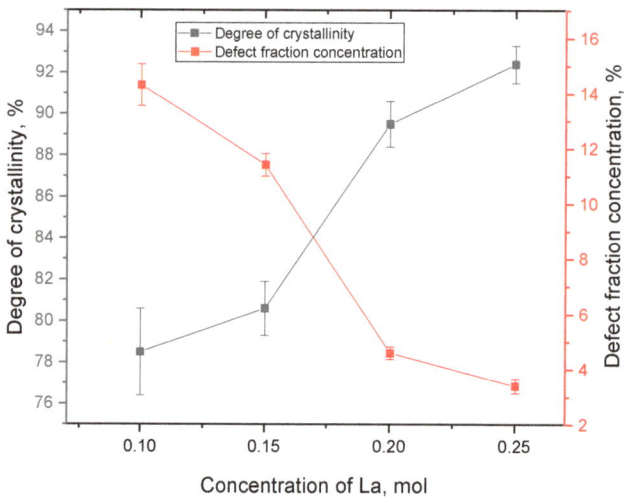

Figure 5. Results of changes in the crystallinity degree and the defect fraction concentration in the composition of ceramics with variation of the components.

The obtained results of the change in the phase composition and structural perovskite-like ceramics depending on the concentration of the initial components are in good agreement with previous similar studies of strontium lanthanum ferrites obtained by various methods. Thus, for example, the results of the change in the phase composition with an increase in lanthanum concentration with the formation of the $Sr_2Fe_2O_5$ phase with the brownmillerite structure, as well as the displacement of the $LaSr_2Fe_3O_8$ phase and its replacement by the $(La_{0.3}Sr_{0.7})_2FeO_4$ phase, are in good agreement with the results of studies [21,22] in which similar ceramics were obtained by the sol-gel method from strontium, iron, and lanthanum nitrates annealed at a temperature of 400–1300 °C. However, in contrast to works [21–23], these phase transformation processes occur at lower sintering temperatures (1000 °C), and it also makes it possible to control both size effects and phase concentration. In comparison with single-phase ceramics [24], the results of the degree of crystallinity are much higher for two-phase ceramics, which can be further used to increase the strength and stability of ceramics.

Figure 6 shows the results of cyclic voltammetry of the studied ceramic samples depending on the change in the lanthanum concentration in the composition. As can be seen from the data presented, in the case of the dominance of $(La_{0.3}Sr_{0.7})_2FeO_4$ and $LaSr_2Fe_3O_8$ in the composition of phases, the nature of the change in the current–voltage curves is described by a straight line, which has a close to ohmic nature of the change in the conductive properties. At the same time, a change in the lanthanum concentration from 0.10 to 0.15 mol does not lead to significant changes in the current–voltage characteristics, as well as a change in the conductivity nature. In turn, the $LaSr_2Fe_3O_8$ phase displacement with the subsequent formation of an orthorhombic $Sr_2Fe_2O_5$ phase leads to an increase in the slope of the current–voltage curve, as well as the appearance of hints of a transition from an ohmic nature of the conductivity to a semiconductor one. In the case of the complete displacement of the $LaSr_2Fe_3O_8$ phase at lanthanum concentration of 0.25 mol with the formation of two-phase $(La_{0.3}Sr_{0.7})_2FeO_4/Sr_2Fe_2O_5$ type ceramics, a sharp change in the conductivity nature is observed, with the dominance of the semiconductor nature in the region from −4 to 4 V, as well as a decrease by more than 2–3 orders of magnitude of the resistivity. Such a change in the conductive properties of ceramics may be due to the fact that the presence of the LaSr$_2$Fe$_3$O$_8$ phase, which has a dielectric nature, leads to an increase in resistance, while its displacement with the subsequent formation of the Sr$_2$Fe$_2$O$_5$ phase leads to a change not only in the resistance value, but also in the conductivity nature from ohmic to semiconductor. Such a change in the conductivity type with a change in the phase composition of perovskite-like ceramics, followed by the LaSr$_2$Fe$_3$O$_8$ phase displacement and the Sr$_2$Fe$_2$O$_5$ phase formation, indicates that ionic conductivity begins to predominate in the structure.

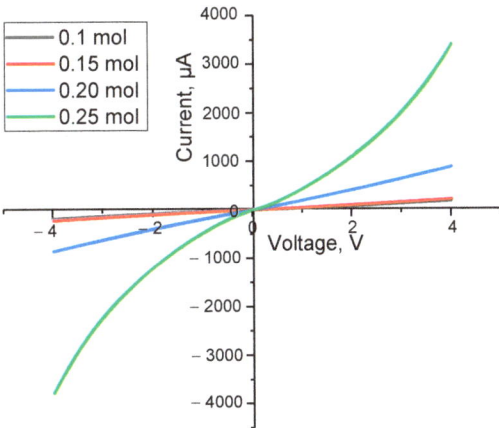

Figure 6. Graphs of cyclic current–voltage characteristics of synthesized ceramics depending on the lanthanum concentration.

Figure 7 shows the results of changes in the electrochemical characteristics of the synthesized ceramics depending on the lanthanum concentration when they are used as an air electrode of SOFC element at a temperature of 550 °C. The fabrication of an SOFC cell was carried out according to the standard technology for obtaining three-layer devices, where Sm$_{0.2}$Ce$_{0.8}$O$_{2-\delta}$ powder at a concentration of 0.20 g pressed with synthesized ceramics and nickel foam was used as an electrolyte, and synthesized ceramics and nickel foam (Ni$_{0.8}$Co$_{0.15}$Al$_{0.05}$LiO$_{2-\delta}$) were used as an air electrode. Nickel foam was also used as an anode. After pressing, the resulting devices were annealed at a temperature of 600 °C for 4 h in an argon atmosphere to compact the SOFC. The performance of the cells was demonstrated in a humidified hydrogen fuel 3% H$_2$O and air with atmospheric oxidizers, at a drift rate of H$_2$/air of 100 ± 5% mL/min.

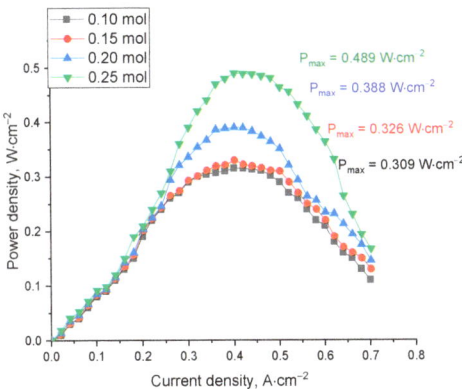

Figure 7. Electrochemical characteristics: current density–specific power (I–P) characteristic curves.

As can be seen from the data presented, the maximum value of specific power is achieved at an open circuit voltage of approximately 1.1–1.15 V; it varies from 0.309 W·cm^{-2} to 0.489 W·cm^{-2} depending on the phase composition of the ceramics. At the same time, in the case when the $(La_{0.3}Sr_{0.7})_2FeO_4$ phase dominates in the composition of the ceramics, the power value changes insignificantly from 0.309 to 0.326 W·cm^{-2} with a decrease in the $LaSr_2Fe_3O_8$ phase contribution from 31 to 22%. Formation of the $Sr_2Fe_2O_5$ phase in the composition of ceramics leads to an increase in P_{max} to 0.388 W·cm^{-2}, and subsequent displacement of the $LaSr_2Fe_3O_8$ phase leads to an increase in P_{max} to 0.489 W·cm^{-2}, which, in comparison with the values for samples of the $(La_{0.3}Sr_{0.7})_2FeO_4/LaSr_2Fe_3O_8$ phase composition, is an increase in power by more than 50%.

Figure 8 shows the Cole–Cole plot and the frequency dependences of the complex permittivity plotted for the obtained samples. Figure 8a shows that the dependences $Z''(Z')$ for samples with 0.10, 0.15, and 0.20 do not represent a semicircle characteristic of a parallel RC chain [24]. This can be explained by the influence of polarization at the ceramic/electrode interface, as well as by a more complex electrical equivalent circuit describing the structure of the resulting ceramic. The data from the Cole–Cole diagrams correlate with the measurements of I-V dependences and electrochemical characteristics, since, as shown in the I-V and I-P dependences, with an increase in the mole fraction of $La(NO_3)_3$, the electrical power increases, the electrical resistance decreases, and the value of the impedance Z', Z'' decreases [25]. The frequency dependences of the dielectric permittivity show (Figure 8b) that in the frequency range of 2–200 kHz, the values of the real part and the imaginary part of the permittivity decrease significantly with increasing electric field frequencies. In the low-frequency region, the processes of interfacial polarization significantly increase the values of ε', ε'', since the electric charge has time to fully accumulate at the semiconductor/dielectric interface. Such interfaces most likely arise due to the pronounced semiconductor properties of the $(La_{0.3}Sr_{0.7})FeO_4$ phase.

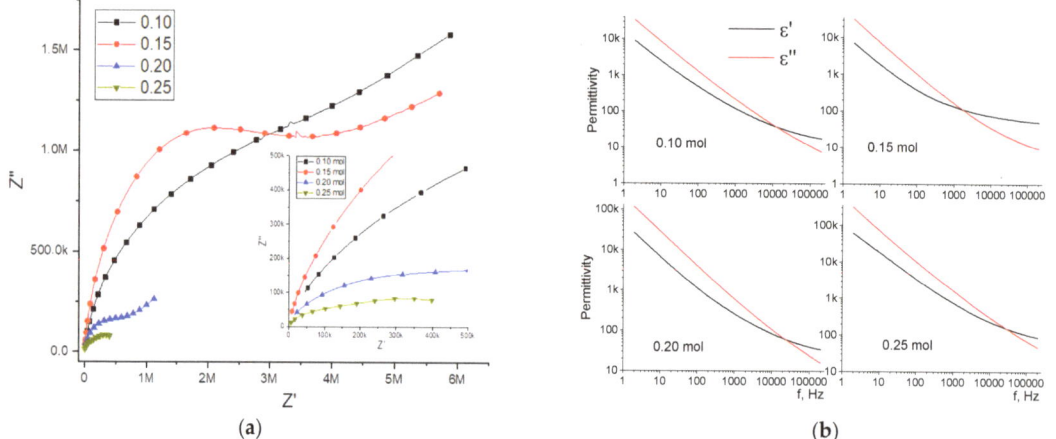

Figure 8. Cole–Cole diagrams $Z''(Z')$ (**a**) and frequency dependences of the complex permittivity (**b**).

For example, it was shown in [26] that the system $La_{1-x}Nd_xSrFeO_4$ with the space group m has a resistivity of $77.62 - 4 \times 10^2$ Ω·cm for the values x = 0.0, 0.3, 0.6, 0.9. Conductivity in $La_{1-x}Nd_xSrFeO_4$ is explained by the mechanism of hopping polaron electrical conductivity [26] with a low activation energy of the process 0.2–0.3 eV. The decrease in electrical resistance is associated with a decrease in the antiferromagnetic interaction when Fe^{3+} cations are replaced by Dy^{3+} cations, which leads to an increase in electron mobility. Semiconductor properties of $LaSrFeO_4$ are also reported in other works, where the electrical resistivity is in the range of 2.4×10^3–10^7 Ohm·cm [27–29]. $LaSrFeO_4$ with an excess of Sr creates holes and p-type conduction in Fe^{4+}/Fe^{3+} pairs when La^{3+} is replaced by the Sr^{2+}

cation. In the case of oxygen deficiency, Fe^{2+}/Fe^{3+} pairs appear in ceramics, generating electrons and n-type conductivity. From Table 3, it is evident that the electrical resistivity ρ_{DC} of the obtained samples is higher than in the above-mentioned works [27–30]. This is due to the presence of a phase with high electrical resistance. The $LaSr_2Fe_3O_8$ phase in the synthesized ceramics acts as an insulator.

Table 3. Comparison of permittivity ε', dielectric loss tangent $\tan\delta = \varepsilon''/\varepsilon'$ at 10 Hz, 10,000 Hz and DC resistance ρ_{DC}.

Mol.	ε' (10 Hz)	ε' (10,000 Hz)	tan δ (10 Hz)	tan δ (10,000 Hz)	$\rho_{DC} \times 10^7$, Ω·cm	Conductivity, S/cm
0.10	2461 ± 50	40.7 ± 2.0	3.4 ± 0.2	1.1 ± 0.1	12.9 ± 0.4	0.07
0.15	1848.6 ± 80	73.7 ± 5.0	4.0 ± 0.3	0.5 ± 0.1	13.1 ± 0.4	0.07
0.20	6727 ± 110	84.1 ± 6.0	4.0 ± 0.3	1.3 ± 0.2	2.9 ± 0.2	0.34
0.25	18,382 ± 120	231.6 ± 11.0	4.3 ± 0.2	1.2 ± 0.3	2.4 ± 0.3	0.41

From the analysis of the I-V dependences, it can be seen that in the samples obtained, the highest current value is observed where the $Sr_2Fe_2O_5$ phase is present and the $LaSr_2Fe_3O_8$ phase is absent. This allows us to conclude that the electrical resistivity of $Sr_2Fe_2O_5$ is the lowest of all the phases that make up ceramics, and its presence significantly reduces the value of the electrical resistance. This is also evidenced by the fact that the shape of the I-V curve changes from a linear dependence to an exponential dependence characteristic of a metal/semiconductor structure. Among the scientific publications for the $Sr_2Fe_2O_5$ compound, data on the electrical properties are scarce, but a low electrical resistance is reported, for example, in [31]. The presence of this component largely determines the dielectric properties of the resulting ceramics. The effective value of the permittivity, conductivity, electrical resistivity, and loss tangent are determined by the concentration of the phases that make up the ceramic.

From Table 3, which compares the dielectric parameters at fixed frequencies and DC conductivity, it can be seen that the largest values of ε' and the smallest values of ρ_{DC} are characteristic of samples with the presence of the $Sr_2Fe_2O_5$ phase.

Table 3 presents the results of the change in the conductivity value for the studied ceramics, obtained by evaluating the Cole–Cole diagrams $Z''(Z')$, which reflect the following. The displacement of the $LaSr_2Fe_3O_8$ phase from the structure with the subsequent formation of the $Sr_2Fe_2O_5$ phase in the structure leads to an increase in specific conductivity by approximately an order of magnitude, which indicates a change in the nature of conductivity in ceramics, associated not only with a change in the phase composition, but also with a change in the charge transfer mechanisms.

4. Conclusions

The paper presents the results of changes in the structural, conductive, and electrochemical properties of perovskite-like ceramics based on lanthanum strontium ferrite, depending on the variation of the components. An analysis of morphological tests showed that an increase in the lanthanum concentration leads to a change in the morphology and grain sizes associated with their crushing and the formation of close-packed ceramics. In this case, as shown by X-ray phase analysis, the change in morphology is due to the processes of phase transformations of the $(La_{0.3}Sr_{0.7})_2FeO_4/LaSr_2Fe_3O_8 \rightarrow (La_{0.3}Sr_{0.7})_2FeO_4/LaSr_2Fe_3O_8/Sr_2Fe_2O_5 \rightarrow (La_{0.3}Sr_{0.7})_2FeO_4/Sr_2Fe_2O_5$ type. The results of impedance spectroscopy showed that with an increase in the concentration of lanthanum from 0.10 to 0.25 mol in synthesized ceramics, the dielectric constant value increases significantly from 40.72 to 231.69, the dielectric loss tangent increases from 1.07 to 1.29 at a frequency of 10,000 Hz, and there is a decrease in specific electrical resistance from 1.29×10^8 to 2.37×10^7 Ω·cm. These changes in the dielectric characteristics are associated with phase transformations in ceramics, in which the ratios between phases with different electrical characteristics change. An analysis of electrochemical measurements showed that the displacement of the $LaSr_2Fe_3O_8$ phase with the subsequent formation of the $Sr_2Fe_2O_5$ phase in the structure leads to an increase in the specific power by more than 50% compared to two-phase ceramics containing the $LaSr_2Fe_3O_8$ phase.

Author Contributions: Conceptualization, D.I.S., D.B.B., A.T.Z., R.I.S. and A.L.K.; methodology, M.V.Z. and A.L.K.; formal analysis, M.V.Z., R.I.S., D.I.S. and A.L.K.; investigation, M.V.Z., D.I.S. and A.L.K.; resources, A.L.K.; writing—original draft preparation, review, and editing, D.B.B., R.I.S., D.I.S. and A.L.K.; visualization, A.L.K.; supervision, A.L.K. All authors have read and agreed to the published version of the manuscript.

Funding: This research was funded by the Science Committee of the Ministry of Education and Science of the Republic of Kazakhstan (No. AP13068071).

Institutional Review Board Statement: Not applicable.

Informed Consent Statement: Not applicable.

Data Availability Statement: Not applicable.

Conflicts of Interest: The authors declare that they have no conflict of interest.

References

1. Krainova, D.; Saetova, N.; Farlenkov, A.; Khodimchuk, A.; Polyakova, I.; Kuzmin, A. Long-term stability of SOFC glass sealant under oxidising and reducing atmospheres. *Ceram. Int.* **2021**, *47*, 8973–8979. [CrossRef]
2. Jin, Y.; Sheng, J.; Hao, G.; Guo, M.; Hao, W.; Yang, Z.; Xiong, X.; Peng, S. Highly dense $(Mn, Co)_3O_4$ spinel protective coating derived from Mn–Co metal precursors for SOFC interconnect applications. *Int. J. Hydrog. Energy* **2022**, *47*, 13960–13968. [CrossRef]
3. Qu, P.; Xiong, D.; Zhu, Z.; Gong, Z.; Li, Y.; Li, Y.; Fan, L.; Liu, Z.; Wang, P.; Liu, C.; et al. Inkjet printing additively manufactured multilayer SOFCs using high quality ceramic inks for performance enhancement. *Addit. Manuf.* **2021**, *48*, 102394. [CrossRef]
4. Garai, M.; Singh, S.P.; Karmakar, B. Mica $(KMg_3AlSi_3O_{10}F_2)$ based glass-ceramic composite sealant with thermal stability for SOFC application. *Int. J. Hydrog. Energy* **2020**, *46*, 23480–23488. [CrossRef]
5. Ullmann, H.; Trofimenko, N.; Tietz, F.; Stöver, D.; Ahmad-Khanlou, A. Correlation between thermal expansion and oxide ion transport in mixed conducting perovskite-type oxides for SOFC cathodes. *Solid State Ion.* **2000**, *138*, 79–90. [CrossRef]
6. Bamburov, A.D.; Markov, A.A.; Patrakeev, M.V.; Leonidov, I.A. The impact of Ba substitution in lanthanum-strontium ferrite on the mobility of charge carriers. *Solid State Ion.* **2019**, *332*, 86–92. [CrossRef]
7. Sažinas, R.; Andersen, K.B.; Hansen, K.K. Facilitating oxygen reduction by silver nanoparticles on lanthanum strontium ferrite cathode. *J. Solid State Electrochem.* **2020**, *24*, 609–621. [CrossRef]
8. Götsch, T.; Köpfle, N.; Schlicker, L.; Carbonio, E.A.; Hävecker, M.; Knop-Gericke, A.; Schloegl, R.; Bekheet, M.F.; Gurlo, A.; Doran, A. Treading in the limited stability regime of lanthanum strontium ferrite—Reduction, phase change and exsolution. *ECS Trans.* **2019**, *91*, 1771. [CrossRef]
9. Yang, J.; Chen, L.; Cai, D.; Zhang, H.; Wang, J.; Guan, W. Study on the strontium segregation behavior of lanthanum strontium cobalt ferrite electrode under compression. *Int. J. Hydrog. Energy* **2021**, *46*, 9730–9740. [CrossRef]
10. Cao, Z.; Zhang, Y.; Miao, J.; Wang, Z.; Lü, Z.; Sui, Y.; Huang, X.; Jiang, W. Titanium-substituted lanthanum strontium ferrite as a novel electrode material for symmetrical solid oxide fuel cell. *Int. J. Hydrog. Energy* **2015**, *40*, 16572–16577. [CrossRef]
11. Chen, Y.; Galinsky, N.; Wang, Z.; Li, F. Investigation of perovskite supported composite oxides for chemical looping conversion of syngas. *Fuel* **2014**, *134*, 521–530. [CrossRef]
12. Marcucci, A.; Zurlo, F.; Sora, I.N.; Placidi, E.; Casciardi, S.; Licoccia, S.; Di Bartolomeo, E. A redox stable Pd-doped perovskite for SOFC applications. *J. Mater. Chem. A* **2019**, *7*, 5344–5352. [CrossRef]
13. Striker, T.; Ruud, J.A.; Gao, Y.; Heward, W.J.; Steinbruchel, C. A-site deficiency, phase purity and crystal structure in lanthanum strontium ferrite powders. *Solid State Ion.* **2007**, *178*, 1326–1336. [CrossRef]
14. Tang, Y.; Chiabrera, F.; Morata, A.; Cavallaro, A.; Liedke, M.O.; Avireddy, H.; Maller, M.; Butterling, M.; Wagner, A.; Stchakovsky, M.; et al. Ion Intercalation in Lanthanum Strontium Ferrite for Aqueous Electrochemical Energy Storage Devices. *ACS Appl. Mater. Interfaces* **2022**, *14*, 18486–18497. [CrossRef] [PubMed]
15. Götsch, T.; Köpfle, N.; Grünbacher, M.; Bernardi, J.; Carbonio, E.A.; Hävecker, M.; Knop-Gericke, A.; Bekheet, M.F.; Schlicker, L.; Doran, A.; et al. Crystallographic and electronic evolution of lanthanum strontium ferrite ($La_{0.6}Sr_{0.4}FeO_{3-\delta}$) thin film and bulk model systems during iron exsolution. *Phys. Chem. Chem. Phys.* **2019**, *21*, 3781–3794. [CrossRef]
16. Jiang, S.P. Development of lanthanum strontium cobalt ferrite perovskite electrodes of solid oxide fuel cells—A review. *Int. J. Hydrog. Energy* **2019**, *44*, 7448–7493. [CrossRef]
17. Wang, J.; Fu, L.; Yang, J.; Wu, K.; Zhou, J.; Wu, K. Cerium and ruthenium co-doped $La_{0.7}Sr_{0.3}FeO_{3-\delta}$ as a high-efficiency electrode for symmetrical solid oxide fuel cell. *J. Rare Earths* **2021**, *39*, 1095–1099. [CrossRef]

18. Thalinger, R.; Gocyla, M.; Heggen, M.; Klötzer, B.; Penner, S. Exsolution of Fe and SrO nanorods and nanoparticles from lanthanum strontium ferrite $La_{0.6}Sr_{0.4}FeO_{3-\delta}$ materials by hydrogen reduction. *J. Phys. Chem. C* **2015**, *119*, 22050–22056. [CrossRef]
19. Deka, D.J.; Gunduz, S.; Fitzgerald, T.; Miller, J.T.; Co, A.C.; Ozkan, U.S. Production of syngas with controllable H_2/CO ratio by high temperature co-electrolysis of CO_2 and H_2O over Ni and Co-doped lanthanum strontium ferrite perovskite cathodes. *Appl. Catal. B Environ.* **2019**, *248*, 487–503. [CrossRef]
20. Zhou, Q.; Yuan, C.; Han, D.; Luo, T.; Li, J.; Zhan, Z. Evaluation of $LaSr_2Fe_2CrO_{9-\delta}$ as a Potential Electrode for Symmetrical Solid Oxide Fuel Cells. *Electrochim. Acta* **2014**, *133*, 453–458. [CrossRef]
21. Sedykh, V.D.; Rybchenko, O.G.; Nekrasov, A.N.; Koneva, I.E.; Kulakov, V.I. Effect of the Oxygen Content on the Local Environment of Fe Atoms in Anion-Deficient $SrFeO_{3-\delta}$. *Phys. Solid State* **2019**, *61*, 1099–1106. [CrossRef]
22. Sedykh, V.D.; Rybchenko, O.G.; Barkovskii, N.V.; Ivanov, A.I.; Kulakov, V.I. Structural Features of a Substituted Strontium Ferrite $Sr_{1-x}La_xFeO_{3-\delta}$: I. $Sr_2LaFe_3O_{9-\delta}$. *Phys. Solid State* **2021**, *63*, 1775–1784. [CrossRef]
23. Sedykh, V.D.; Rybchenko, O.G.; Barkovskii, N.V.; Ivanov, A.I.; Kulakov, V.I. Structural Transformations and Valence States of Fe in Substituted Strontium Ferrite $Sr_2LaFe_3O_{9-\delta}$. *J. Surf. Investig. X-Ray Synchrotron Neutron Tech.* **2021**, *15*, 1138–1143. [CrossRef]
24. Leonidov, I.A.; Kozhevnikov, V.L.; Patrakeev, M.V.; Mitberg, E.B.; Poeppelmeier, K.R. High-temperature electrical conductivity of $Sr_{0.7}La_{0.3}FeO_{3-\delta}$. *Solid State Ion.* **2001**, *144*, 361–369. [CrossRef]
25. Asandulesa, M.; Kostromin, S.; Tameev, A.; Aleksandrov, A.; Bronnikov, S. Molecular Dynamics and Conductivity of a PTB7:PC71BM Photovoltaic Polymer Blend: A Dielectric Spectroscopy Study. *ACS Appl. Polym. Mater.* **2021**, *3*, 4869–4878. [CrossRef]
26. Barsoukov, E.; Macdonald, J.R. (Eds.) *Impedance Spectroscopy Theory, Experiment, and Applications*, 2nd ed.; John Wiley &Sons, Inc.: Hoboken, NJ, USA, 2005; pp. 1–40.
27. King, H.W.; Castelliz, K.M.; Murphy, G.J. Crystal structure and electrical resistivity of ceramics with compositions La_2MO_4, $La_{1.5}Sr_{0.5}MO_4$, and $LaSrMO_4$. *J. Can. Ceram. Soc.* **1987**, *55*, 10–14.
28. Omata, T.; Ueda, K.; Ueda, N.; Katada, M.; Fujitsu, S.; Hashimoto, T.; Kawazoe, H. Preparation of oxygen excess $SrLaFeO_{4+\delta}$ and its electrical and magnetic properties. *Solid State Commun.* **1993**, *88*, 807–811. [CrossRef]
29. Fujihara, S.; Nakata, T.; Kozuka, H.; Yoko, T. The Effects of Substitution of Alkaline Earths or Y for La on Structure and Electrical Properties of $LaSrFeO_4$. *J. Solid State Chem.* **1995**, *115*, 456–463. [CrossRef]
30. Singh, D.; Singh, S.; Mahajan, A.; Choudhary, N. Effect of substitution of magnetic rare earth Nd at non-magnetic La site on structure and properties of $LaSrFeO_4$. *Ceram. Int.* **2014**, *40*, 1183–1188. [CrossRef]
31. Monteiro Filho, F.; Rocha, W.A.; Correa, R.R.; Almeida, R.M.; Paschoal, C.W.A. Characterization of $Sr_2Fe_2O_5$ ferrite. In Proceedings of the 53 Brazilian Congress on Ceramics, Guaruja, Brazil, 7–10 June 2009; Volume 12, p. 53.

Article

Analysis of Electromagnetic Effects on Vibration of Functionally Graded GPLs Reinforced Piezoelectromagnetic Plates on an Elastic Substrate

Mohammed Sobhy [1,2,*] and F. H. H. Al Mukahal [1]

[1] Department of Mathematics and Statistics, College of Science, King Faisal University, P.O. Box 400, Al-Ahsa 31982, Saudi Arabia; falmukahal@kfu.edu.sa
[2] Department of Mathematics, Faculty of Science, Kafrelsheikh University, Kafrelsheikh 33516, Egypt
* Correspondence: msobhy@kfu.edu.sa

Abstract: A new nanocomposite piezoelectromagnetic plate model is developed for studying free vibration based on a refined shear deformation theory (RDPT). The present model is composed of piezoelectromagnetic material reinforced with functionally graded graphene platelets (FG-GPLs). The nanocomposite panel rests on Winkler–Pasternak foundation and is subjected to external electric and magnetic potentials. It is assumed that the electric and magnetic properties of the GPLs are proportional to those of the electromagnetic materials. The effective material properties of the plate are estimated based on the modified Halpin–Tsai model. A refined graded rule is introduced to govern the variation in the volume fraction of graphene through the thickness of the plate. The basic partial differential equations are provided based on Hamilton's principle and then solved analytically to obtain the eigenfrequency for different boundary conditions. To check the accuracy of the present formulations, the depicted results are compared with the published ones. Moreover, impacts of the variation in elastic foundation stiffness, plate geometry, electric potential, magnetic potential, boundary conditions and GPLs weight fraction on the vibration of the smart plate are detailed and discussed.

Keywords: electric potential; magnetic potential; refined four-unknown shear deformation theory; functionally graded graphene nanosheets; piezoelectromagnetic materials; free vibration

Citation: Sobhy, M.; Al Mukahal, F.H.H. Analysis of Electromagnetic Effects on Vibration of Functionally Graded GPLs Reinforced Piezoelectromagnetic Plates on an Elastic Substrate. *Crystals* **2022**, *12*, 487. https://doi.org/10.3390/cryst12040487

Academic Editor: Shujun Zhang

Received: 22 February 2022
Accepted: 29 March 2022
Published: 31 March 2022

Publisher's Note: MDPI stays neutral with regard to jurisdictional claims in published maps and institutional affiliations.

Copyright: © 2022 by the authors. Licensee MDPI, Basel, Switzerland. This article is an open access article distributed under the terms and conditions of the Creative Commons Attribution (CC BY) license (https://creativecommons.org/licenses/by/4.0/).

1. Introduction

Due to the efficiency of piezoelectric and piezoelectromagnetic materials in converting electro-mechanical and magnetic energies to each other, such materials arise in a wide range of fields and industries including heat exchangers, smart devices, nuclear devices and electromechanical systems [1–12]. Moreover, such materials have a key role in nano-electro-mechanical systems such as sensors, actuators, nanogenerators, active controllers and nano-robotics [13]. Generally, nanocomposite structures reinforced with graphene are the main component of these devices since they are well-known for their stunning electro-mechanical properties. As a consequence of this, a large number of academic works have been devoted to investigate the properties of such materials. Based on the theory of linear piezoelectromagneticity, Hu and Li [8] obtained the expressions for singular stresses, electric displacements and magnetic fields in a piezoelectromagnetic plate with a Griffith crack subjected to longitudinal shear loads. Moreover, Ke and Wang [14] determined the linear natural frequencies of size-dependent electromagnetic nanobeams under external electric, magnetic potentials and a uniform temperature field based upon the nonlocal-Timoshenko beam theory and nonlocal elasticity theory. In another piece of research, Ke et al. [15] studied the free vibration behavior of magneto-electro-elastic (MEE) nanoplates employing the nonlocal and Kirchhoff plate theories. Employing the same mentioned nanobeams hypotheses, the bending, buckling and free vibration of MEE composites have been analyzed by Li et al. [16], taking into consideration the small-scale-dependent coefficients and the strength of the electric and induced magnetic fields on the transverse displacement, rotation, buckling loads and natural frequency. Exact solutions for anisotropic functionally graded and multi-layer MEE rectangular plates with material properties varying exponentially along the thickness direction were proposed by Pan and Han [17]. Farajpour et al. [18] depicted a nonlocal plate model to consider the size-effect on the nonlinear vibration behavior of MEE composite nanoplates subjected to external electromagnetic loading conditions. Furthermore, Farajpour et al. [19] considered the higher order deformations along with the higher and lower order nonlocal effects for the nonlinear bucking of orthotropic nanoplates in a thermal environment. Jamalpoor et al. [20] determined the closed-form solutions for natural frequencies and mechancial buckling loads of double-MEE nanoplate systems exposed to initial external electric voltage and magnetic potentials embedded in a viscoelastic medium. Mehditabar et al. [21] studied 3D magnetothermoelastic responses of FGM cylindrical shells using differential quadrature (DQ) technique subjected to non-uniform internal pressure. Under the influence of external electric voltage, Zenkour and Aljadani [22] determined the electro-mechanical buckling behavior of simply supported rectangular functionally-graded piezoelectric (FGP) plates using a quasi-3D refined plate theory. The impacts of various parameters, such as magnetic parameters, electrical potentials, rotation speed, thickness-to-radios ratio, and axial and circumferential wave numbers, on the free vibration of the rotating FG polymer cylindrical shells integrated by two piezo-electromagnetic (PEM) face sheets were evaluated by Meskini and Ghasemi [23]. Abazid and Sobhy [24] explained the thermal and EM size-dependent bending of simply-supported FG piezoelectric (FGP) microplates embedded on a Pasternak elastic foundation depending on a novel refined four-variable shear deformation plate theory, with the help of modified couple-stress theory. Nanoplate problems subjected to hygrothermal loads have been proposed in [25,26], using different nonlocal theories. Chen et al. [27] showed analytical formulations for the wave propagation studies in MEE multilayered plates with nonlocal properties, and selected two types of sandwich plates to investigate the nonlocal parameter on the dispersion curve. The impact of shear deformation and angular velocity on the wave propagation behaviours of MEE rotary nanobeams is performed by Ebrahimi and Dabbagh [28], using the nonlocal strain gradient theory (NSGT). Furthermore, the thermomechanical buckling, free vibration and wave propagation in smart piezoelectromagnetic nanoplates in a hygrothermal medium embedded in an elastic substrate was explored by Abazid [29]. In addition, some useful studies of sandwich structures in various configurations of FGPMs or FGPs, and subjected to various loadings, have also been performed by some eminent researchers [2,30–32].

Graphene reveals exceptionally superior electromechanical and physical properties (Potts et al. [33]), and is composed of a single thick layer of sp^2 joined carbon atoms arranged in a 2D hexagonal form. Graphene is the most powerful material that has been detected, since its tensile strength equals about 130.5 GPa, it has a Young's modulus greater than 1 TPa, a mass of 1 m^2 is 0.77 mg and electrical conductivity 1000 times greater than copper for electric current-carrying capacity (Papageorgiou et al. [34]). Furthermore, the specific-surface-area of the graphene is exactly 2630 m^2/g (Papageorgiou et al. [34]), whereas that of carbon nanotubes is in the range of 100–1000 m^2/g. Graphene has been investigated as an ideal effective reinforcement of the piezoelectric material composite structures due to it enhances their electromechanical features and stiffness (see, e.g., Yang et al. [35], Forsat et al. [36], Khorasani et al. [11], Thai et al. [37,38] and Phung-Van et al. [39]). Mao et al. [40] presented the vibrational characteristics of the FG piezoelectric composite micro-plate reinforced with graphene nano-sheets (GNSs) upon the non-local constitutive relation and von-Karman geometric non-linearity, in which the equations of motion were solved via the differential quadrature (DQ) method. They showed that the concentration of graphene nano-platelets, exterior voltage, nonlocal parameters, geometrical and piezoelectric properties of the GNSs, as well as the elasticity parameters of the Winkler elastic foundation, has a key insight in the linear and nonlinear dynamic behaviors of the GNSs reinforcing FG piezoelectric composite micro-plates. Sobhy [41] developed an analytical method for the ME-thermal bending of FG-GNSs reinforced composite doubly-curved shallow shells surrounded by two smart face sheets of PEM with several boundary conditions. The obtained results of Mao and Zhang [42] showed that the piezoconductive characteristics of the GNSs nanofillers can significantly improve the stiffness of the FG-GNSs plates. Mao and Zhang [43] investigated the post-buckling and buckling properties of FG-GNSs plates subjected to electric potential and mechanical loads, in which the equations of motion were obtained by the combination of differential quadrature method and direct iterative technique. Furthermore, Sobhy et al. [44] investigated the change in thermal buckling in PFG-GNSs beams exposed to external electric voltage in a humid environment. In addition, some experimental works have shown that graphene reinforcements can obviously enhance the mechanical properties of the piezoelectric. Abolhasani et al. [45] investigated the influence of graphene reinforced PVDF (polyvinylidene fluoride) composites on the morphology, crystallinity, polymorphism and electrical outputs, on which the enhanced PVDF/graphene can be a potential application for portable self-powering devices. Xu et al.'s experiments [46] showed the positive piezoconductive impact in suspended graphene layers, which strongly depends on the layer-number.

As viewed in the above survey, many papers have been performed to study the behavior of GPL reinforced piezoelectric plates. However, the GPL reinforced piezoelectromagnetic (PEM) plate has not been considered in the literature. Motivated by this deficiency and to fill this gap, the present article is conducted to provide a comprehensive and clear scientific vision for the free vibration of piezoelectromagnetic plates reinforced with functionally graded graphene nanosheets (FG-GNSs) under simply supported conditions. In addition, a refined four-variable shear deformation theory is introduced to define the displacement field. The present nanocomposite panel is assumed to be resting on a Winkler–Pasternak foundation and subjected to external electric and magnetic potentials. The electric and magnetic properties of the GNSs are assumed to be proportional to those of the electromagnetic plate. Moreover, in accordance with the modified Halpin–Tsai model, the effective material properties of the plate are evaluated. To govern the variation in the volume fraction of graphene through the thickness of the plate, a refined graded rule is used. Hamilton's principle is employed effectively to deduce equations of the motion mathematically, and then solved analytically to obtain the eigenfrequency. In order to validate the present formulations, the depicted results are compared with the published ones. Furthermore, the detailed parametric investigation of the variation in various parameters on the vibration of the smart plate are discussed.

2. Plate Configuration

In this paper, we consider multi-layered piezoelectromagnetic plates reinforced with functionally graded graphene platelets with rectangular platform in which the length, width and total thickness are indicated by L, W and h, respectively, as displayed in Figure 1. The Cartesian coordinates system (x, y, z) is used to prescribe the infinitesimal deformations of the plate. This plate is assumed to be resting on a Winkler–Pasternak elastic substrate with stiffness $(\hat{J}_1, \hat{J}_2)$, and subjected to an external electric voltage and magnetic potential.

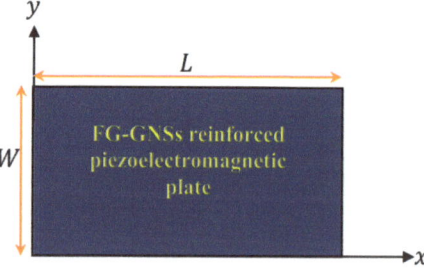

Figure 1. Geometry and coordinates of a naocomposite piezoelectromagnetic plate.

Within the framework of the modified Halpin–Tsai model, the effective Young's modulus $E^{(i)}$ for the ith layer of the composite plate is given by [47–49]

$$E^{(i)} = \frac{E_{pzm}}{8} \left[\frac{3\left(1 + 2\kappa_1 \delta_1^{GP} V^{(i)GP}\right)}{1 - \kappa_1 V^{(i)GP}} + \frac{5\left(1 + 2\kappa_2 \delta_2^{GP} V^{(i)GP}\right)}{1 - \kappa_2 V^{(i)GP}} \right], \qquad (1)$$

where E_{pzm} is the piezoelectromagnetic Young's modulus, $\delta_1^{GP} = L^{GP}/h^{GP}$ and $\delta_2^{GP} = W^{GP}/h^{GP}$, in which L^{GP}, W^{GP} and h^{GP} stand for the length, width and thickness of GNSs, respectively; $V^{(i)GP}$ denotes the volume fraction of graphene for the ith layer, and the coefficients κ_1 and κ_2 are displayed as

$$\kappa_1 = \frac{E^{GP} - E_{pzm}}{E^{GP} + 2\delta_1^{GP} E_{pzm}}, \qquad \kappa_2 = \frac{E^{GP} - E_{pzm}}{E^{GP} + 2\delta_2^{GP} E_{pzm}}, \qquad (2)$$

in which E^{GP} indicates Young's modulus of the GNSs. Moreover, the effective material properties of the FG-GNSs reinforced composite plate for the ith layer, namely Poisson's ration ν^i and mass density ρ^i, can be evaluated as

$$\begin{aligned} \nu^i &= V^{(i)GP} \nu^{GP} + \left(1 - V^{(i)GP}\right) \nu_{pzm}, \\ \rho^i &= V^{(i)GP} \rho^{GP} + \left(1 - V^{(i)GP}\right) \rho_{pzm}, \end{aligned} \qquad (3)$$

in which ν^{GP} (ν_{pzm}) and ρ^{GP} (ρ_{pzm}) are Poisson's ratio of the GNSs (piezoelectromagnetic) and the mass density of the GNSs (piezoelectromagnetic), respectively. With respect to a modified piece-wise rule, the volume fraction of the GNSs will be varied across the thickness of the plate layers, as presented in Figure 2, and in the present analysis, four different cases of FG-GNSs distribution are considered as follows:

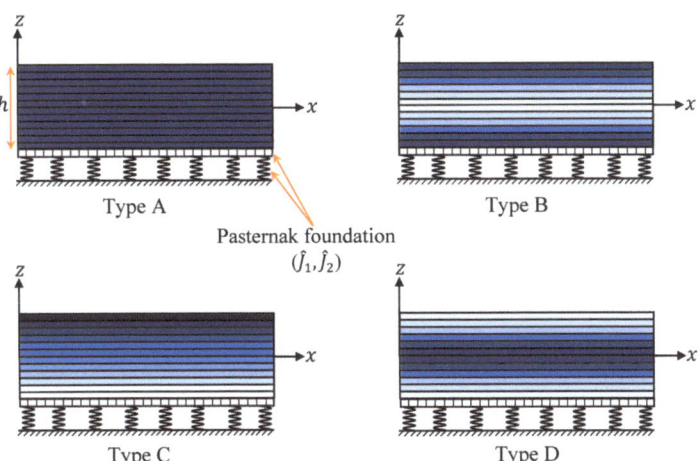

Figure 2. Various types of naocomposite piezoelectromagnetic plates.

2.1. Type A

In the present case, the graphene platelets may be uniformly distributed throughout thickness h. As a result, the volume fraction of the ith layer can be written by

$$V^{(i)GP} = V_{max}^{GP} = \frac{\rho_{pzm} W_G}{\rho_{pzm} W_G + \rho^{GP}(1 - W_G)}, \quad \text{for UD} \qquad (4)$$

in which W_G stands for the weight fraction of GNSs.

2.2. Type B

The volume fraction of graphene has a minimum value at the mid-plane of the sheets. While it equals a maximum value at the upper and lower sheets. Therefore, $V^{(i)GP}$ is written as

$$V^{(i)GP} = V_{max}^{GP} \left(\frac{|2i - N - 1| - 1}{N - 2} \right)^k \quad \text{for FG-X,} \tag{5}$$

in which N represents the number of layers of the plate, and $0 \leq k \leq \infty$ is a power-law index.

2.3. Type C

In this particular case, the volume fraction of graphene decreases in a monotonic pattern from its maximum value V_{max}^{GP} at the top-surface, to a minimum value at the bottom-surface of the plate. We can denote the gradient-distributions of graphene by FG-V. Therefore, the volume fraction for each layer can be written as

$$V^{(i)GP} = V_{max}^{GP} \left(\frac{i-1}{N-1} \right)^k \quad \text{for FG-V.} \tag{6}$$

2.4. Type D

In this state, the volume fraction of graphene equals zero at the top- and bottom-plate surfaces. It has a maximum value at the mid-plane of the plate. As a result, $V^{(i)GP}$ can be expressed as

$$V^{(i)GP} = V_{max}^{GP} \left(\frac{|2i - N - 1| + 1 - N}{2 - N} \right)^k, \quad \text{for FG-O.} \tag{7}$$

3. Constitutive Equations

A modified refined four-variable shear deformation plate theory ([50,51]), that is enhanced by introducing the bending and shear effects, is presented to describe the displacement components given by

$$\begin{Bmatrix} U(x,y,z) \\ V(x,y,z) \\ W(x,y,z) \end{Bmatrix} = \begin{Bmatrix} u_0(x,y) \\ v_0(x,y) \\ w_b(x,y) + w_s(x,y) \end{Bmatrix} - z \begin{Bmatrix} w_{b,x} \\ w_{b,y} \\ 0 \end{Bmatrix} - f(z) \begin{Bmatrix} w_{s,x} \\ w_{s,y} \\ 0 \end{Bmatrix} \tag{8}$$

where $u_0(x,y)$ and $v_0(x,y)$ denote the component of the displacement of the mid-plane along the x- and y-axes, respectively. In accordance with the theory of Shimpi ([50]), the transverse deflection $W(x,y,z)$ is divided into two components $w_b(x,y)$ and $w_s(x,y)$ which indicate the bending and shear displacements, respectively, and $f(z) = z - g(z)$. In addition, in the above mentioned theory, a new-shape function is introduced which represents the configuration of the shear stress through the thickness of the plate, and it can be declared as

$$g(z) = \frac{z}{1 + (z/h)^2} - \frac{5}{8} \frac{z^3}{h^2}. \tag{9}$$

The definition of the shape function $g(z)$ allows the recovery of a variety of theories in the literature, such as the third-order theory (TDPT) [52] $g(z) = z - 4z^3/3h^2$, the sinusoidal theory (SDPT) [53] $g(z) = (h/\pi)\sin(z\pi/h)$, and the exponential theory (EDPT) [54] $g(z) = ze^{-2z^2/h^2}$. Based on the displacement field (8), one can obtain the components of the strains as:

$$\left\{\begin{array}{c}\varepsilon_{xx}\\ \varepsilon_{yy}\\ \varepsilon_{xy}\end{array}\right\}=\left\{\begin{array}{c}\varepsilon_{xx}^{(0)}\\ \varepsilon_{yy}^{(0)}\\ \varepsilon_{xy}^{(0)}\end{array}\right\}+z\left\{\begin{array}{c}\varepsilon_{xx}^{(1)}\\ \varepsilon_{yy}^{(1)}\\ \varepsilon_{xy}^{(1)}\end{array}\right\}+f(z)\left\{\begin{array}{c}\varepsilon_{xx}^{(2)}\\ \varepsilon_{yy}^{(2)}\\ \varepsilon_{xy}^{(2)}\end{array}\right\},$$

(10)

$$\left\{\begin{array}{c}\varepsilon_{xz}\\ \varepsilon_{yz}\end{array}\right\}=g'\left\{\begin{array}{c}\varepsilon_{xz}^{(2)}\\ \varepsilon_{yz}^{(2)}\end{array}\right\},\quad g'=\frac{dg}{dz},$$

where

$$\left\{\begin{array}{c}\varepsilon_{xx}^{(0)}\\ \varepsilon_{yy}^{(0)}\\ \varepsilon_{xy}^{(0)}\end{array}\right\}=\left\{\begin{array}{c}u_{0,x}\\ v_{0,y}\\ u_{0,y}+v_{0,x}\end{array}\right\},\quad\left\{\begin{array}{c}\varepsilon_{xx}^{(1)}\\ \varepsilon_{yy}^{(1)}\\ \varepsilon_{xy}^{(1)}\end{array}\right\}=-\left\{\begin{array}{c}w_{b,xx}\\ w_{b,yy}\\ 2w_{b,xy}\end{array}\right\},$$

$$\left\{\begin{array}{c}\varepsilon_{xx}^{(2)}\\ \varepsilon_{yy}^{(2)}\\ \varepsilon_{xy}^{(2)}\end{array}\right\}=-\left\{\begin{array}{c}w_{s,xx}\\ w_{s,yy}\\ 2w_{s,xy}\end{array}\right\},\quad\left\{\begin{array}{c}\varepsilon_{xz}^{(2)}\\ \varepsilon_{yz}^{(2)}\end{array}\right\}=\left\{\begin{array}{c}w_{s,x}\\ w_{s,y}\end{array}\right\}.$$

(11)

In respect to the piezoelasticity theory [2,14,41,55], the constitutive relation for the components of the stresses σ can be expressed as follows

$$\left\{\begin{array}{c}\sigma_{xx}\\ \sigma_{yy}\\ \sigma_{yz}\\ \sigma_{xz}\\ \sigma_{xy}\end{array}\right\}^{(i)}=\begin{bmatrix}C_{11}&C_{12}&0&0&0\\ C_{12}&C_{22}&0&0&0\\ 0&0&C_{44}&0&0\\ 0&0&0&C_{55}&0\\ 0&0&0&0&C_{66}\end{bmatrix}^{(i)}\left\{\begin{array}{c}\varepsilon_{xx}\\ \varepsilon_{yy}\\ \varepsilon_{yz}\\ \varepsilon_{xz}\\ \varepsilon_{xy}\end{array}\right\}-\begin{bmatrix}0&0&\zeta_{13}\\ 0&0&\zeta_{23}\\ 0&\zeta_{24}&0\\ \zeta_{15}&0&0\\ 0&0&0\end{bmatrix}^{(i)}\left\{\begin{array}{c}\hat{E}_{1}\\ \hat{E}_{2}\\ \hat{E}_{3}\end{array}\right\}$$

$$-\begin{bmatrix}0&0&\eta_{13}\\ 0&0&\eta_{23}\\ 0&\eta_{24}&0\\ \eta_{15}&0&0\\ 0&0&0\end{bmatrix}^{(i)}\left\{\begin{array}{c}M_{1}\\ M_{2}\\ M_{3}\end{array}\right\}$$

(12)

where $(C_{11}, C_{12}, C_{22}, C_{44}, C_{55}, C_{66})$ denote the elastic coefficients of the FG-GNSs composite plate which are expressed as:

$$C_{11}^{(i)}=C_{22}^{(i)}=\frac{E^{(i)}}{1-\left(v^{(i)}\right)^{2}},\quad C_{12}^{(i)}=\frac{v^{(i)}E^{(i)}}{1-\left(v^{(i)}\right)^{2}},\quad C_{44}^{(i)}=C_{55}^{(i)}=C_{66}^{(i)}=\frac{E^{(i)}}{2(1+v^{(i)})}.\quad(13)$$

Furthermore, the remaining coefficients of piezoelectromagnetic and dielectric $(\zeta_{13}, \zeta_{23}, \zeta_{24}, \zeta_{15})$ and $(\eta_{13}, \eta_{23}, \eta_{24}, \eta_{15})$, respectively, can be calculated as follows:

$$\zeta_{jk}^{(i)}=V^{(i)GP}\zeta_{jk}^{GP}+\left(1-V^{(i)GP}\right)\zeta_{jk,pzm},$$
$$\eta_{jk}^{(i)}=V^{(i)GP}\eta_{jk}^{GP}+\left(1-V^{(i)GP}\right)\eta_{jk,pzm},\quad j,k=1,2,\ldots,5,$$

(14)

where ζ_{jk}^{GP} ($\zeta_{jk,pzm}$) and η_{jk}^{GP} ($\eta_{jk,pzm}$) denote the piezoelectric coefficients of GNSs (piezoelectromagnetic), and the dielectric coefficients of GNSs (piezoelectromagnetic), respectively. The electric displacements D_i and the magnetic induction B_i are written as

$$
\begin{Bmatrix} D_x \\ D_y \\ D_z \end{Bmatrix}^{(i)} = \begin{bmatrix} 0 & 0 & 0 & \zeta_{15} & 0 \\ 0 & 0 & \zeta_{24} & 0 & 0 \\ \zeta_{13} & \zeta_{23} & 0 & 0 & 0 \end{bmatrix}^{(i)} \begin{Bmatrix} \varepsilon_{xx} \\ \varepsilon_{yy} \\ \varepsilon_{yz} \\ \varepsilon_{xz} \\ \varepsilon_{xy} \end{Bmatrix} + \begin{bmatrix} f_{11} & 0 & 0 \\ 0 & f_{22} & 0 \\ 0 & 0 & f_{33} \end{bmatrix}^{(i)} \begin{Bmatrix} \hat{E}_1 \\ \hat{E}_2 \\ \hat{E}_3 \end{Bmatrix}
$$

$$
+ \begin{bmatrix} g_{11} & 0 & 0 \\ 0 & g_{22} & 0 \\ 0 & 0 & g_{33} \end{bmatrix}^{(i)} \begin{Bmatrix} M_1 \\ M_2 \\ M_3 \end{Bmatrix},
$$
(15)

$$
\begin{Bmatrix} B_x \\ B_y \\ B_z \end{Bmatrix}^{(i)} = \begin{bmatrix} 0 & 0 & 0 & \eta_{15} & 0 \\ 0 & 0 & \eta_{24} & 0 & 0 \\ \eta_{13} & \eta_{23} & 0 & 0 & 0 \end{bmatrix}^{(i)} \begin{Bmatrix} \varepsilon_{xx} \\ \varepsilon_{yy} \\ \varepsilon_{yz} \\ \varepsilon_{xz} \\ \varepsilon_{xy} \end{Bmatrix} + \begin{bmatrix} g_{11} & 0 & 0 \\ 0 & g_{22} & 0 \\ 0 & 0 & g_{33} \end{bmatrix}^{(i)} \begin{Bmatrix} \hat{E}_1 \\ \hat{E}_2 \\ \hat{E}_3 \end{Bmatrix}
$$

$$
+ \begin{bmatrix} r_{11} & 0 & 0 \\ 0 & r_{22} & 0 \\ 0 & 0 & r_{33} \end{bmatrix}^{(i)} \begin{Bmatrix} M_1 \\ M_2 \\ M_3 \end{Bmatrix},
$$
(16)

where (f_{11}, f_{22}, f_{33}), (g_{11}, g_{22}, g_{33}), and (r_{11}, r_{22}, r_{33}) denote, respectively, the dielectric, magneto-electric and magnetic constants of the FG-GNSs composite plate, which can be defined as follows

$$
\begin{aligned}
f_{jj}^{(i)} &= V^{(i)GP} f_{jj}^{GP} + \left(1 - V^{(i)GP}\right) f_{jj,pzm} \\
g_{jj}^{(i)} &= V^{(i)GP} g_{jj}^{GP} + \left(1 - V^{(i)GP}\right) g_{jj,pzm}, \quad j = 1, 2, 3 \\
r_{jj}^{(i)} &= V^{(i)GP} r_{jj}^{GP} + \left(1 - V^{(i)GP}\right) r_{jj,pzm},
\end{aligned}
$$
(17)

The electric field $\hat{E}$ and magnetic field M can take the following form [14,41]:

$$
\hat{E} = -\nabla \Phi, \quad M = -\nabla \Psi
$$
(18)

where Φ and Ψ represent, respectively, the electric and magnetic potentials of the FG-GNSs composite plate that can defined as [14,41]:

$$
\begin{aligned}
\Phi(x, y, z) &= -\phi(x, y) \cos\left(\frac{\pi z}{h}\right) + \frac{2z\hat{V}_0}{h}, \\
\Psi(x, y, z) &= -\psi(x, y) \cos\left(\frac{\pi z}{h}\right) + \frac{2z\hat{P}_0}{h},
\end{aligned}
$$
(19)

in which $\hat{V}_0$ and $\hat{P}_0$ are, respectively, the external applied electric and magnetic potentials, $\phi(x, y)$ and $\psi(x, y)$ denote the electric and magnetic potentials of the middle surface of the plate. Substituting Equation (19) into Equation (18) gives the components of the electric and magnetic fields

$$
\begin{Bmatrix} \hat{E}_1 \\ \hat{E}_2 \\ \hat{E}_3 \end{Bmatrix} = \begin{Bmatrix} \phi_{,x} \cos\left(\frac{\pi z}{h}\right) \\ \phi_{,y} \cos\left(\frac{\pi z}{h}\right) \\ -\frac{\pi}{h} \phi \sin\left(\frac{\pi z}{h}\right) \end{Bmatrix} - \begin{Bmatrix} 0 \\ 0 \\ \frac{2\hat{V}_0}{h} \end{Bmatrix} \quad \begin{Bmatrix} M_1 \\ M_2 \\ M_3 \end{Bmatrix} = \begin{Bmatrix} \psi_{,x} \cos\left(\frac{\pi z}{h}\right) \\ \psi_{,y} \cos\left(\frac{\pi z}{h}\right) \\ -\frac{\pi}{h} \psi \sin\left(\frac{\pi z}{h}\right) \end{Bmatrix} - \begin{Bmatrix} 0 \\ 0 \\ \frac{2\hat{P}_0}{h} \end{Bmatrix}.
$$
(20)

4. Governing Equations

In order to establish the equations of motion associated with the displacement field in Equation (8), the variation in the strain energy $\delta \mathcal{K}_S$, the kinetic energy $\delta \mathcal{K}_K$, the work done by the in-plane piezoelectromagnetic load $\mathcal{K}_{EM}$ and the elastic foundations $\delta \mathcal{K}_F$ can be stated by Hamilton's principle

$$\int_0^t \left(\delta \mathcal{K}_S + \delta \mathcal{K}_K - \delta \mathcal{K}_{EM} - \delta \mathcal{K}_F \right) dt = 0, \quad (21)$$

where

$$\delta \mathcal{K}_S = \sum_{i=1}^{N} \int_V \left(\sigma_{jk}^{(i)} \delta \varepsilon_{jk} - D_j^{(i)} \delta \hat{E}_j - B_j^{(i)} \delta M_j \right) dV,$$

$$\delta \mathcal{K}_K = \sum_{i=1}^{N} \int_V \rho^{(i)} \left(\ddot{u} \delta u + \ddot{v} \delta v + \ddot{w} \delta w \right) dV, \quad (22)$$

$$\delta \mathcal{K}_{EM} = \int_A \left(N^E + N^M \right) \left(\frac{\partial^2 w}{\partial x^2} + \frac{\partial^2 w}{\partial y^2} \right) \delta w \, dA,$$

$$\delta \mathcal{K}_F = -\int_A \left(\hat{J}_1 w - \hat{J}_2 \nabla^2 w \right) \delta w \, dA,$$

in which $\hat{J}_1$ and $\hat{J}_2$ denote, respectively, the springs and shear layer foundation stiffness, N^E and N^M denote the in-plane hygrothermal forces which are defined, respectively, as

$$N^E = \sum_{i=1}^{N} \int_{h_i}^{h_{i+1}} \xi_{13}^{(i)} \frac{2\hat{V}_0}{h} dz, \quad N^M = \sum_{i=1}^{N} \int_{h_i}^{h_{i+1}} \eta_{13}^{(i)} \frac{2\hat{P}_0}{h} dz. \quad (23)$$

Substituting ε_{jk}, $\hat{E}_j$ and M_j into the first equation of Equation (22) gives

$$\delta \mathcal{K}_S = \sum_{i=1}^{N} \int_V \left[\sigma_{xx}^{(i)} \delta \left(\varepsilon_{xx}^{(0)} + z \varepsilon_{xx}^{(1)} + f \varepsilon_{xx}^{(2)} \right) + \sigma_{yy}^{(i)} \delta \left(\varepsilon_{yy}^{(0)} + z \varepsilon_{yy}^{(1)} + f \varepsilon_{yy}^{(2)} \right) \right.$$
$$+ \sigma_{xy}^{(i)} \delta \left(\varepsilon_{xy}^{(0)} + z \varepsilon_{xy}^{(1)} + f \varepsilon_{xy}^{(2)} \right) + \sigma_{xz}^{(i)} \delta \left(g' \varepsilon_{xz}^{(2)} \right) + \sigma_{yz}^{(i)} \delta \left(g' \varepsilon_{yz}^{(2)} \right) - D_1^{(i)} \delta \left(\frac{\partial \phi}{\partial x} \cos\left(\frac{\pi z}{h}\right) \right)$$
$$- D_2^{(i)} \delta \left(\frac{\partial \phi}{\partial y} \cos\left(\frac{\pi z}{h}\right) \right) + D_3^{(i)} \delta \left(\frac{\pi \phi}{h} \sin\left(\frac{\pi z}{h}\right) \right) - B_1^{(i)} \delta \left(\frac{\partial \psi}{\partial x} \cos\left(\frac{\pi z}{h}\right) \right)$$
$$\left. - B_2^{(i)} \delta \left(\frac{\partial \psi}{\partial y} \cos\left(\frac{\pi z}{h}\right) \right) + B_3^{(i)} \delta \left(\frac{\pi \psi}{h} \sin\left(\frac{\pi z}{h}\right) \right) \right] dV, \quad (24)$$

or

$$\delta \mathcal{K}_S = \int_A \left[\hat{N}_{xx} \delta \varepsilon_{xx}^{(0)} + \hat{M}_{xx} \delta \varepsilon_{xx}^{(1)} + S_{xx} \delta \varepsilon_{xx}^{(2)} + \hat{N}_{yy} \delta \varepsilon_{yy}^{(0)} + \hat{M}_{yy} \delta \varepsilon_{yy}^{(1)} + S_{yy} \delta \varepsilon_{yy}^{(2)} \right.$$
$$+ \hat{N}_{xy} \delta \varepsilon_{xy}^{(0)} + \hat{M}_{xy} \delta \varepsilon_{xy}^{(1)} + S_{xy} \delta \varepsilon_{xy}^{(2)} + Q_{xz} \delta \varepsilon_{xz}^{(2)} + Q_{yz} \delta \varepsilon_{yz}^{(2)} - R_1 \delta \frac{\partial \phi}{\partial x} - R_2 \delta \frac{\partial \phi}{\partial y} \quad (25)$$
$$\left. + R_3 \delta \phi - T_1 \delta \frac{\partial \psi}{\partial x} - T_2 \delta \frac{\partial \psi}{\partial y} + T_3 \delta \psi \right] dA,$$

where

$$\{\hat{N}_{xx}, \hat{M}_{xx}, S_{xx}\} = \sum_{i=1}^{N} \int_{h_i}^{h_{i+1}} \sigma_{xx}^{(i)} \{1, z, f\} \, dz,$$

$$\{\hat{N}_{yy}, \hat{M}_{yy}, S_{yy}\} = \sum_{i=1}^{N} \int_{h_i}^{h_{i+1}} \sigma_{yy}^{(i)} \{1, z, f\} \, dz, \quad (26)$$

$$\{\hat{N}_{xy}, \hat{M}_{xy}, S_{xy}\} = \sum_{i=1}^{N} \int_{h_i}^{h_{i+1}} \sigma_{xy}^{(i)} \{1, z, f\} \, dz,$$

$$\{Q_{xz}, Q_{yz}\} = \sum_{i=1}^{N} \int_{h_i}^{h_{i+1}} g' \{\sigma_{xz}^{(i)}, \sigma_{yz}^{(i)}\} \, dz,$$

$$R_1 = \sum_{i=1}^{N} \int_{h_i}^{h_{i+1}} D_1^{(i)} \cos\left(\frac{\pi z}{h}\right) dz, \quad R_2 = \sum_{i=1}^{N} \int_{h_i}^{h_{i+1}} D_2^{(i)} \cos\left(\frac{\pi z}{h}\right) dz,$$

$$R_3 = \sum_{i=1}^{N} \int_{h_i}^{h_{i+1}} D_3^{(i)} \frac{\pi}{h} \sin\left(\frac{\pi z}{h}\right) dz, \quad T_1 = \sum_{i=1}^{N} \int_{h_i}^{h_{i+1}} B_1^{(i)} \cos\left(\frac{\pi z}{h}\right) dz, \qquad (27)$$

$$T_2 = \sum_{i=1}^{N} \int_{h_i}^{h_{i+1}} B_2^{(i)} \cos\left(\frac{\pi z}{h}\right) dz, \quad T_3 = \sum_{i=1}^{N} \int_{h_i}^{h_{i+1}} B_3^{(i)} \frac{\pi}{h} \sin\left(\frac{\pi z}{h}\right) dz,$$

By substituting $\sigma_{ij}^{(i)}$, D_i and B_i into Equations (26) and (27), we can obtain

$$\begin{Bmatrix} \hat{N}_{11} \\ \hat{N}_{22} \\ \hat{M}_{11} \\ \hat{M}_{22} \\ S_{11} \\ S_{22} \\ R_3 \\ T_3 \end{Bmatrix} = \begin{bmatrix} a_{11} & a_{12} & a_{13} & b_{11} & b_{12} & b_{13} \\ b_{11} & b_{12} & b_{13} & b_{21} & b_{22} & b_{23} \\ a_{12} & a_{32} & a_{33} & b_{12} & b_{32} & b_{33} \\ b_{12} & b_{32} & b_{33} & b_{22} & b_{42} & b_{43} \\ a_{13} & a_{33} & a_{53} & b_{13} & b_{33} & b_{53} \\ b_{13} & b_{33} & b_{53} & b_{23} & b_{43} & b_{63} \\ d_{11} & d_{31} & d_{51} & d_{21} & d_{41} & d_{61} \\ d_{12} & d_{32} & d_{52} & d_{22} & d_{42} & d_{62} \end{bmatrix} \begin{Bmatrix} \varepsilon_{11}^{(0)} \\ \varepsilon_{11}^{(1)} \\ \varepsilon_{11}^{(2)} \\ \varepsilon_{22}^{(0)} \\ \varepsilon_{22}^{(1)} \\ \varepsilon_{22}^{(2)} \end{Bmatrix} + \begin{Bmatrix} d_{11} \\ d_{21} \\ d_{31} \\ d_{41} \\ d_{51} \\ d_{61} \\ \bar{d}_{11} \\ \bar{d}_{12} \end{Bmatrix} \phi + \begin{Bmatrix} d_{12} \\ d_{22} \\ d_{32} \\ d_{42} \\ d_{52} \\ d_{62} \\ \bar{d}_{12} \\ \bar{d}_{22} \end{Bmatrix} \psi + \begin{Bmatrix} N_1^E + N_1^M \\ N_2^E + N_2^M \\ \hat{M}_1^E + \hat{M}_1^M \\ \hat{M}_2^E + \hat{M}_2^M \\ S_1^E + S_1^M \\ S_2^E + S_2^M \\ R_3^E + R_3^M \\ T_3^E + T_3^M \end{Bmatrix} \qquad (28)$$

$$\begin{Bmatrix} \hat{N}_{12} \\ \hat{M}_{12} \\ S_{12} \end{Bmatrix} = \begin{bmatrix} \bar{a}_{31} & \bar{a}_{32} & \bar{a}_{33} \\ \bar{a}_{32} & \bar{a}_{42} & \bar{a}_{43} \\ \bar{a}_{33} & \bar{a}_{43} & \bar{a}_{53} \end{bmatrix} \begin{Bmatrix} \varepsilon_{12}^{(0)} \\ \varepsilon_{12}^{(1)} \\ \varepsilon_{12}^{(2)} \end{Bmatrix} \qquad (29)$$

$$\begin{Bmatrix} Q_{13} \\ R_1 \\ T_1 \end{Bmatrix} = \begin{bmatrix} \hat{a}_{11} & -\hat{a}_{12}^E & -\hat{a}_{13}^M \\ \hat{a}_{12}^E & \hat{a}_{13}^E & \hat{a}_{14}^E \\ \hat{a}_{13}^M & \hat{a}_{14}^E & \hat{a}_{33}^M \end{bmatrix} \begin{Bmatrix} \varepsilon_{13}^{(2)} \\ \phi_{,x} \\ \psi_{,x} \end{Bmatrix}$$

$$\begin{Bmatrix} Q_{23} \\ R_2 \\ T_2 \end{Bmatrix} = \begin{bmatrix} \hat{a}_{21} & -\hat{a}_{22}^E & -\hat{a}_{23}^M \\ \hat{a}_{22}^E & \hat{a}_{23}^E & \hat{a}_{24}^E \\ \hat{a}_{23}^M & \hat{a}_{24}^E & \hat{a}_{43}^M \end{bmatrix} \begin{Bmatrix} \varepsilon_{23}^{(2)} \\ \phi_{,y} \\ \psi_{,y} \end{Bmatrix} \qquad (30)$$

where

$$\{a_{11}, a_{12}, a_{13}, a_{32}, a_{33}, a_{53}\} = \sum_{i=1}^{N} \int_{h_i}^{h_{i+1}} C_{11}^{(i)} \{1, z, f, z^2, zf, f^2\} dz,$$

$$\{b_{11}, b_{12}, b_{13}, b_{32}, b_{33}, b_{53}\} = \sum_{i=1}^{N} \int_{h_i}^{h_{i+1}} C_{12}^{(i)} \{1, z, f, z^2, zf, f^2\} dz,$$

$$\{b_{21}, b_{22}, b_{23}, b_{42}, b_{43}, b_{63}\} = \sum_{i=1}^{N} \int_{h_i}^{h_{i+1}} C_{22}^{(i)} \{1, z, f, z^2, zf, f^2\} dz,$$

$$\begin{Bmatrix} d_{11}, d_{12} \\ d_{31}, d_{32} \\ d_{51}, d_{52} \end{Bmatrix} = \sum_{i=1}^{N} \int_{h_i}^{h_{i+1}} \frac{\pi}{h} \sin\left(\frac{\pi z}{h}\right) \begin{Bmatrix} 1 \\ z \\ f \end{Bmatrix} \{\zeta_{13}^{(i)}, \eta_{13}^{(i)}\} dz, \qquad (31)$$

$$\begin{Bmatrix} d_{21}, d_{22} \\ d_{41}, d_{42} \\ d_{61}, d_{62} \end{Bmatrix} = \sum_{i=1}^{N} \int_{h_i}^{h_{i+1}} \frac{\pi}{h} \sin\left(\frac{\pi z}{h}\right) \begin{Bmatrix} 1 \\ z \\ f \end{Bmatrix} \{\zeta_{23}^{(i)}, \eta_{23}^{(i)}\} dz,$$

$$\{\bar{d}_{11}, \bar{d}_{12}, \bar{d}_{13}\} = -\sum_{i=1}^{N} \int_{h_i}^{h_{i+1}} \left(\frac{\pi}{h}\right)^2 \sin^2\left(\frac{\pi z}{h}\right) \{f_{33}^{(i)}, g_{33}^{(i)}, r_{33}^{(i)}\} dz,$$

$$\begin{Bmatrix} N_2^E \\ \widehat{M}_2^E \\ S_2^E \end{Bmatrix} = \sum_{i=1}^{N} \int_{h_i}^{h_{i+1}} \left(\frac{2\hat{V}_0}{h}\right) \begin{Bmatrix} 1 \\ z \\ f \end{Bmatrix} \zeta_{23}^{(i)} dz,$$

$$\begin{Bmatrix} N_2^M, \\ \widehat{M}_2^M \\ S_2^M \end{Bmatrix} = \sum_{i=1}^{N} \int_{h_i}^{h_{i+1}} \left(\frac{2\hat{P}_0}{h}\right) \begin{Bmatrix} 1 \\ z \\ f \end{Bmatrix} \eta_{23}^{(i)} dz,$$

$$\begin{Bmatrix} \widehat{M}_1^E \\ S_1^E \end{Bmatrix} = \sum_{i=1}^{N} \int_{h_i}^{h_{i+1}} \left(\frac{2\hat{V}_0}{h}\right) \begin{Bmatrix} z \\ f \end{Bmatrix} \zeta_{13}^{(i)} dz, \qquad (32)$$

$$\begin{Bmatrix} \widehat{M}_1^M \\ S_1^M \end{Bmatrix} = \sum_{i=1}^{N} \int_{h_i}^{h_{i+1}} \left(\frac{2\hat{P}_0}{h}\right) \begin{Bmatrix} z \\ f \end{Bmatrix} \eta_{13}^{(i)} dz,$$

$$\{R_3^E, T_3^E\} = -\sum_{i=1}^{N} \int_{h_i}^{h_{i+1}} \{f_{33}^{(i)}, g_{33}^{(i)}\} \left(\frac{2\hat{V}_0}{h}\right) \left(\frac{\pi}{h}\right) \sin\left(\frac{\pi z}{h}\right) dz,$$

$$\{R_3^M, T_3^M\} = -\sum_{i=1}^{N} \int_{h_i}^{h_{i+1}} \{g_{33}^{(i)}, r_{33}^{(i)}\} \left(\frac{2\hat{P}_0}{h}\right) \left(\frac{\pi}{h}\right) \sin\left(\frac{\pi z}{h}\right) dz,$$

$$\{\bar{a}_{31}, \bar{a}_{32}, \bar{a}_{33}, \bar{a}_{42}, \bar{a}_{43}, \bar{a}_{53}\} = \sum_{i=1}^{N} \int_{h_i}^{h_{i+1}} C_{66}^{(i)} \{1, z, f, z^2, zf, f^2\} dz,$$

$$\{\hat{a}_{11}, \hat{a}_{21}\} = \sum_{i=1}^{N} \int_{h_i}^{h_{i+1}} \{C_{55}^{(i)}, C_{44}^{(i)}\} g'^2 dz, \qquad (33)$$

$$\{\hat{a}_{12}^E, \hat{a}_{22}^E, \hat{a}_{13}^M, \hat{a}_{23}^M\} = \sum_{i=1}^{N} \int_{h_i}^{h_{i+1}} g' \cos\left(\frac{\pi z}{h}\right) \{\zeta_{15}^{(i)}, \zeta_{24}^{(i)}, \eta_{15}^{(i)}, \eta_{24}^{(i)}\} dz,$$

$$\{\hat{a}_{13}^E, \hat{a}_{14}^E, \hat{a}_{33}^M, \hat{a}_{23}^E, \hat{a}_{24}^E, \hat{a}_{43}^M\} = \sum_{i=1}^{N} \int_{h_i}^{h_{i+1}} \{f_{11}^{(i)}, g_{11}^{(i)}, r_{11}^{(i)}, f_{22}^{(i)}, g_{22}^{(i)}, r_{22}^{(i)}\} \cos^2\left(\frac{\pi z}{h}\right) dz,$$

Substituting Equations (22) and (25) into Equation (21) leads to the equation of motion as follows:

$$\frac{\partial \widehat{N}_{xx}}{\partial x} + \frac{\partial \widehat{N}_{xy}}{\partial y} = \mathcal{I}_{11}\ddot{u}_0 - \mathcal{I}_{12}\frac{\partial \ddot{w}_b}{\partial x} - \mathcal{I}_{13}\frac{\partial \ddot{w}_s}{\partial x},$$

$$\frac{\partial \widehat{N}_{xy}}{\partial x} + \frac{\partial \widehat{N}_{yy}}{\partial y} = \mathcal{I}_{11}\ddot{v}_0 - \mathcal{I}_{12}\frac{\partial \ddot{w}_b}{\partial y} - \mathcal{I}_{13}\frac{\partial \ddot{w}_s}{\partial y},$$

$$\frac{\partial^2 \widehat{M}_{xx}}{\partial x^2} + 2\frac{\partial^2 \widehat{M}_{xy}}{\partial x \partial y} + \frac{\partial^2 \widehat{M}_{yy}}{\partial y^2} - \mathcal{J}_1 w + \mathcal{J}_2 \nabla^2 w + (N^E + N^M)\nabla^2 w = \mathcal{I}_{11}\ddot{w}$$
$$+ \mathcal{I}_{12}\left(\frac{\partial \ddot{u}_0}{\partial x} + \frac{\partial \ddot{v}_0}{\partial y}\right) - \nabla^2 \left(\mathcal{I}_{22}\ddot{w}_b + \mathcal{I}_{23}\ddot{w}_s\right), \qquad (34)$$

$$\frac{\partial^2 S_{xx}}{\partial x^2} + 2\frac{\partial^2 S_{xy}}{\partial x \partial y} + \frac{\partial^2 S_{yy}}{\partial y^2} - \mathcal{J}_1 w + \mathcal{J}_2 \nabla^2 w + (N^E + N^M)\nabla^2 w = \mathcal{I}_{11}\ddot{w}$$
$$+ \mathcal{I}_{13}\left(\frac{\partial \ddot{u}_0}{\partial x} + \frac{\partial \ddot{v}_0}{\partial y}\right) - \nabla^2 \left(\mathcal{I}_{23}\ddot{w}_b + \mathcal{I}_{33}\ddot{w}_s\right),$$

$$\frac{\partial R_1}{\partial x} + \frac{\partial R_2}{\partial y} + R_3 = 0,$$

$$\frac{\partial T_1}{\partial x} + \frac{\partial T_2}{\partial y} + T_3 = 0,$$

where

$$\{\mathcal{I}_{11}, \mathcal{I}_{12}, \mathcal{I}_{13}, \mathcal{I}_{22}, \mathcal{I}_{23}, \mathcal{I}_{33}\} = \sum_{i=1}^{N} \int_{h_i}^{h_{i+1}} \rho^{(i)} \{1, z, f, z^2, zf, f^2\} dz. \qquad (35)$$

5. Solution Procedure

The natural frequency of the FG-GNSs reinforced piezoelectromagnetic rectangular plate is deduced by solving the governing motion Equations (34) with different boundary conditions.

Simply supported (S)

$$v_0 = w_b = w_s = \phi = \psi = \frac{\partial w_b}{\partial y} = \frac{\partial w_s}{\partial y} = \widehat{N}_{xx} = \widehat{M}_{xx} = S_{xx} = 0, \quad \text{at} \quad x = 0, L,$$

$$u_0 = w_b = w_s = \phi = \psi = \frac{\partial w_b}{\partial x} = \frac{\partial w_s}{\partial x} = \widehat{N}_{yy} = \widehat{M}_{yy} = S_{yy} = 0 \quad \text{at} \quad y = 0, W. \quad (36)$$

Clamped (C)

$$u_0 = v_0 = w_i = \phi = \psi = \frac{\partial w_i}{\partial x} = \frac{\partial w_i}{\partial y} = 0, \quad \text{at} \quad x = 0, L, \quad y = 0, W, \quad i = b, s. \quad (37)$$

According to the above boundary conditions, the displacements can be assumed as:

$$u_0 = \sum_{m}^{\infty}\sum_{l}^{\infty} U_{ml} \frac{\partial X_m(x)}{\partial x} \bar{X}_l(y),$$

$$v_0 = \sum_{m}^{\infty}\sum_{l}^{\infty} V_{ml} X_m(x) \frac{\partial \bar{X}_l(y)}{\partial y}, \quad (38)$$

$$\{w_b, w_s, \phi, \psi\} = \sum_{m}^{\infty}\sum_{l}^{\infty} \{\bar{W}_{bml}, \bar{W}_{sml}, \bar{\phi}_{ml}, \bar{\psi}_{ml}\} X_m(x) \bar{X}_l(y),$$

where U_{ml}, V_{ml}, W_{bml}, W_{sml}, $\bar{\phi}_{ml}$ and $\bar{\psi}_{ml}$ are unknown functions. The admissible functions $X_m(x)$ and $\bar{X}_l(y)$ are given in Table 1, noting that $\lambda = m\pi/L$ and $\mu = l\pi/W$.

Table 1. The admissible functions $X_m(x)$ and $\bar{X}_l(y)$.

Boundary Conditions				The Functions	
$x = 0$	$x = L$	$y = 0$	$y = W$	$X_m(x)$	$\bar{X}_l(y)$
S	S	S	S	$\sin(\lambda x)$	$\sin(\mu y)$
C	S	S	S	$\sin(\lambda x)[1 - \cos(\lambda x)]$	$\sin(\mu y)$
C	C	S	S	$\sin^2(\lambda x)$	$\sin(\mu y)$
C	C	C	S	$\sin^2(\lambda x)$	$\sin(\mu y)[1 - \cos(\mu y)]$
C	C	C	C	$\sin^2(\lambda x)$	$\sin^2(\mu y)$

Incorporating Equation (38) into Equations (28)–(30) and then Equation (34) leads to the equations of motion of the nanocomposite piezoelectromagnetic plate as follows:

$$\begin{bmatrix} Y_{11} & Y_{12} & Y_{13} & Y_{14} & Y_{15} & Y_{16} \\ Y_{21} & Y_{22} & Y_{23} & Y_{24} & Y_{25} & Y_{26} \\ Y_{31} & Y_{32} & Y_{33} & Y_{34} & Y_{35} & Y_{36} \\ Y_{41} & Y_{42} & Y_{43} & Y_{44} & Y_{45} & Y_{46} \\ Y_{51} & Y_{52} & Y_{53} & Y_{54} & Y_{55} & Y_{56} \\ Y_{61} & Y_{62} & Y_{63} & Y_{64} & Y_{65} & Y_{66} \end{bmatrix} \begin{Bmatrix} U_{ml} \\ V_{ml} \\ W_{bml} \\ W_{sml} \\ \bar{\phi}_{ml} \\ \bar{\psi}_{ml} \end{Bmatrix} = 0, \quad (39)$$

where the elements of the matrix $[Y]$ are given as:

$$\begin{aligned}
Y_{11} &= a_{11}q_2 + \bar{a}_{31}q_3 + I_{11}q_1\omega^2, \\
Y_{12} &= \bar{b}_{31}q_3 + b_{11}q_3, \\
Y_{13} &= -a_{12}q_2 - I_{12}q_1\omega^2 - 2\bar{b}_{32}q_3 - b_{12}q_3, \\
Y_{14} &= -a_{13}q_2 - I_{13}q_1\omega^2 - 2\bar{b}_{33}q_3 - b_{13}q_3, \\
Y_{15} &= d_{11}q_1, \quad Y_{16} = d_{12}q_1, \\
Y_{21} &= \bar{a}_{31}q_6 + b_{11}q_6, \\
Y_{22} &= \bar{b}_{31}q_6 + a_{11}q_5 + I_{11}q_4\omega^2, \quad Y_{23} = -a_{12}q_5 - I_{12}q_4\omega^2 - 2\bar{b}_{32}q_6 - b_{12}q_6, \\
Y_{24} &= -a_{13}q_5 - I_{13}q_4\omega^2 - 2\bar{b}_{33}q_6 - b_{13}q_6, \quad Y_{25} = d_{11}q_4, \quad Y_{26} = d_{12}q_4, \\
Y_{31} &= a_{12}q_{11} + I_{12}q_9\omega^2 + 2\bar{b}_{32}q_{10} + b_{12}q_{10}, \quad Y_{32} = a_{12}q_{12} + I_{12}q_8\omega^2 + 2\bar{b}_{32}q_{10} + b_{12}q_{10}, \\
Y_{33} &= I_{11}q_7\omega^2 - \hat{J}_1q_7 + N_1^E q_8 + N_1^M q_8 + N_1^E q_9 + N_1^M q_9 - a_{22}q_{11} - a_{22}q_{12} + \hat{J}_2q_8 + \hat{J}_2q_9 \\
&\quad - I_{22}q_8\omega^2 - I_{22}q_9\omega^2 - 4\bar{b}_{42}q_{10} - 2b_{22}q_{10}, \\
Y_{34} &= I_{11}q_7\omega^2 - \hat{J}_1q_7 + N_1^E q_8 + N_1^M q_8 + N_1^E q_9 + N_1^M q_9 - a_{23}q_{11} - a_{23}q_{12} + \hat{J}_2q_8 + \hat{J}_2q_9 \\
&\quad - I_{23}q_8\omega^2 - I_{23}q_9\omega^2 - 4\bar{b}_{43}q_{10} - 2b_{23}q_{10}, \\
Y_{35} &= d_{31}q_8 + d_{31}q_9, \quad Y_{36} = -d_{32}q_8 + d_{32}q_9, \\
Y_{41} &= a_{13}q_{11} + I_{13}q_9\omega^2 + 2\bar{b}_{33}q_{10} + b_{13}q_{10}, \\
Y_{42} &= a_{13}q_{12} + I_{13}q_8\omega^2 + 2\bar{b}_{33}q_{10} + b_{13}q_{10}, \\
Y_{43} &= I_{11}q_7\omega^2 - \hat{J}_1q_7 + N_1^E q_8 + N_1^M q_8 + N_1^E q_9 + N_1^M q_9 - a_{23}q_{11} - a_{23}q_{12} + \hat{J}_2q_8 + \hat{J}_2q_9 \\
&\quad - I_{23}q_8\omega^2 - I_{23}q_9\omega^2 - 4\bar{b}_{43}q_{10} - 2b_{23}q_{10}, \\
Y_{44} &= I_{11}q_7\omega^2 - \hat{J}_1q_7 + N_1^E q_8 + N_1^M q_8 + N_1^E q_9 + N_1^M q_9 - a_{33}q_{11} + \hat{a}_{11}q_8 - a_{33}q_{12} + \hat{a}_{11}q_9 \\
&\quad + \hat{J}_2q_8 + \hat{J}_2q_9 - I_{33}q_8\omega^2 - I_{33}q_9\omega^2 - 4\bar{b}_{53}q_{10} - 2b_{33}q_{10}, \\
Y_{45} &= -\hat{a}_{12}^E q_8 - \hat{a}_{12}^E q_9 + d_{51}q_8 + d_{51}q_9, \\
Y_{46} &= -\hat{a}_{13}^M q_8 - \hat{a}_{13}^M q_9 + d_{52}q_8 + d_{52}q_9, \\
Y_{51} &= d_{11}q_9, \quad Y_{52} = d_{11}q_8, \\
Y_{53} &= -d_{31}q_8 - d_{31}q_9, \\
Y_{54} &= \hat{a}_{12}^E q_8 + \hat{a}_{12}^E q_9 - d_{51}q_8 - d_{51}q_9, \\
Y_{55} &= \hat{a}_{23}^E q_8 + \hat{a}_{23}^E q_9 - \bar{d}_{11}q_7, \\
Y_{56} &= \hat{a}_{14}^E q_8 + \hat{a}_{14}^E q_9 - \bar{d}_{12}q_7, \\
Y_{61} &= d_{12}q_9, \quad Y_{62} = d_{12}q_8, \\
Y_{63} &= -d_{32}q_8 - d_{32}q_9, \\
Y_{64} &= \hat{a}_{13}^M q_8 + \hat{a}_{13}^M q_9 - d_{52}q_8 - d_{52}q_9, \\
Y_{65} &= \hat{a}_{14}^E q_8 + \hat{a}_{14}^E q_9 - \bar{d}_{12}q_7, \\
Y_{66} &= \hat{a}_{33}^M q_8 + \hat{a}_{33}^M q_9 - \bar{d}_{22}q_7.
\end{aligned} \quad (40)$$

where

$$q_1 = \int_0^L \int_0^W \left(\frac{\partial X_m(x)}{\partial x}\bar{X}_l(y)\right)^2 dydx,$$

$$q_2 = \int_0^L \int_0^W \frac{\partial X_m(x)}{\partial x}\frac{\partial^3 X_m(x)}{\partial x^3}(\bar{X}_l(y))^2 dydx,$$

$$q_3 = \int_0^L \int_0^W \left(\frac{\partial X_m(x)}{\partial x}\right)^2 \bar{X}_l(y)\frac{\partial^2 \bar{X}_l(y)}{\partial y^2} dydx,$$

$$q_4 = \int_0^L \int_0^W \left(X_m(x)\frac{\partial \bar{X}_l(y)}{\partial y}\right)^2 dydx,$$

$$q_5 = \int_0^L \int_0^W (X_m(x))^2 \frac{\partial \bar{X}_l(y)}{\partial y}\frac{\partial^3 p\bar{X}_l(y)}{\partial y^3} dydx,$$

$$q_6 = \int_0^L \int_0^W X_m(x)\frac{\partial^2 pX_m(x)}{\partial x^2}\left(\frac{\partial \bar{X}_l(y)}{\partial y}\right)^2 dydx, \qquad (41)$$

$$q_7 = \int_0^L \int_0^W (X_m(x)\bar{X}_l(y))^2 dydx,$$

$$q_8 = \int_0^L \int_0^W (X_m(x))^2 \bar{X}_l(y)\frac{\partial^2 \bar{X}_l(y)}{\partial y^2} dydx,$$

$$q_9 = \int_0^L \int_0^W X_m(x)(\bar{X}_l(y))^2 \frac{\partial^2 pX_m(x)}{\partial x^2} dydx,$$

$$q_{10} = \int_0^L \int_0^W X_m(x)\bar{X}_l(y)\frac{\partial^2 pX_m(x)}{\partial x^2}\frac{\partial^2 \bar{X}_l(y)}{\partial y^2} dydx,$$

$$q_{11} = \int_0^L \int_0^W X_m(x)(\bar{X}_l(y))^2 \frac{\partial^4 pX_m(x)}{\partial x^4} dydx,$$

$$q_{12} = \int_0^L \int_0^W (X_m(x))^2 \bar{X}_l(y)\frac{\partial^4 \bar{X}_l(y)}{\partial y^4} dydx.$$

For a nontrivial solution, the determinante $|Y|$ should be equal to zero. Solving the equation $|Y| = 0$ gives the eigenfrequency of the nanocomposite piezoelectromagnetic plate resting on an elastic substrate.

6. Results and Discussion

In this section, the numerical results of the above formulations are developed for the free vibration of a piezoelectromagnetic plate strengthened by GNSs for different boundary conditions. The material properties of the piezoelectromagnetic material are taken as [14]: $E_{pzm} = 141$ GPa, $\mu_{pzm} = 0.35$, $\rho_{pzm} = 5.55$ g/cm^3, $\zeta_{13,pzm} = \zeta_{23,pzm} = -2.2$ Cm^{-2}, $\zeta_{24,pzm} = \zeta_{15,pzm} = 5.8$ Cm^{-2}, $\eta_{13,pzm} = \eta_{23,pzm} = 290.1$ NA^{-1}m^{-1}, $\eta_{24,pzm} = \eta_{15,pzm} = 275$ NA^{-1}m^{-1}, $f_{11,pzm} = f_{22,pzm} = 5.64 \times 10^{-9}$ CV^{-1}m^{-1}, $f_{33,pzm} = 6.35 \times 10^{-9}$ CV^{-1}m^{-1}, $g_{11,pzm} = g_{22,pzm} = 5.367 \times 10^{-12}$ NsV^{-1}C^{-1}, $g_{33,pzm} = 2737.5 \times 10^{-12}$ NsV^{-1}C^{-1}, $r_{11,pzm} = r_{22,pzm} = -297 \times 10^{-6}$ Ns^{2}C^{-2}, $r_{33,pzm} = 83.5 \times 10^{-6}$ Ns^{2}C^{-2}. While the mechanical properties of graphene are given as [41]: $E^{GP} = 1010$ GPa, $\rho^{GP} = 1.06$ g/cm^3, $\nu^{GP} = 0.186$. In addition, the electromagnetic properties of the graphene nanosheets are assumed to be proportional to that of the piezoelectromagnetic material as: $\zeta_{ij}^{GP} = \chi\zeta_{ij,pzm}$, $\zeta_{ij}^{GP} = \chi\zeta_{ij,pzm}$, $\eta_{ij}^{GP} = \chi\eta_{ij,pzm}$, $f_{ij}^{GP} = \chi f_{ij,pzm}$, $g_{ij}^{GP} = \chi g_{ij,pzm}$, $r_{ij}^{GP} = \chi r_{ij,pzm}$, where χ is the piezoelectromagnetic multiple. The following fixed data have been used (except otherwise stated): $m = l = 1$, $N = 10$, $L^{GP} = 15$ nm, $W^{GP} = 9$ nm, $h^{GP} = 0.188$ nm, $W_G = 0.1$, $L/h = 10$, $L/W = 1$, $k = 1$, $V_0 = 1$, $P_0 = 0.1$, $h = 2$ mm, $\chi = 100$, $J_1 = 100$, $J_2 = 10$. For convenience, the ensuing nondimensional quantities are defined:

$$\Omega = 100h\omega\sqrt{\frac{\rho_{pzm}}{E_{pzm}}}, \quad J_1 = \frac{\hat{J}_1 L^4}{D_{pzm}}, \quad J_2 = \frac{\hat{J}_2 L^2}{D_{pzm}}$$

$$V_0 = \frac{\hat{V}_0 L}{D_{pzm}}, \quad P_0 = \frac{\hat{P}_0 L}{D_{pzm}}, \quad D_{pzm} = \frac{E_{pzm} h^3}{12[1-(\nu_{pzm})^2]}. \tag{42}$$

6.1. Verification

To confirm the accuracy of the current formulations and results, the present outcomes are compared with those depicted by Thai and Choi [56] and Baferani et al. [57]. Three comparison examples are performed, namely for the FG plate without an elastic foundation, resting on a Winkler elastic foundation and resting on a Pasternak elastic foundation as shown in Tables 2–4. In these tables, the nondimensional frequency $\hat{\Omega}$ of Al/Al$_2$O$_3$ (aluminum/alumina) FG plate is presented for different values of the side-to-thickness ratio L/h and the power law index n, and for different shear deformation plate theories. The effective Young's modulus and density are calculated via the following mixture law:

$$E = E_m + (E_c - E_m)\left(\frac{z}{h} + \frac{1}{2}\right)^n, \quad \rho = \rho_m + (\rho_c - \rho_m)\left(\frac{z}{h} + \frac{1}{2}\right)^n, \tag{43}$$

where $E_m = 70$ GPa, $\rho_m = 2.707$ g/cm^3, $E_c = 380$ GPa, $\rho_c = 3.8$ g/cm^3. While Poisson's ratio is given as $\nu = 0.3$. It is noted from Tables 2–4 that the results of the present theory are in a good agreement with the published ones.

Another comparison example is developed here for the natural frequencies $\hat{\Omega}$ of a homogeneous piezoelectromagnetic plate without elastic foundations for different values of the mode numbers as shown in Table 5. The mechanical and electromagnetic properties of the plate are taken as in Refs. [14,29]. Once again, an excellent agreement between the present frequencies and those depicted by Abazid [29] and Ke et al. [14] is also observed.

Table 2. Comparison of nondimensional frequency $\hat{\Omega} = h\omega\sqrt{\rho_m/E_m}$ of FG plate without elastic foundation ($J_1 = 0$, $J_2 = 0$).

n	L/h	Present			Published	
		TDPT	SDPT	RDPT	Ref. [56]	Ref. [57]
0	0.20	0.41543	0.41550	0.41547	0.4154	0.4154
	0.15	0.24543	0.24545	0.24544	0.2454	0.2454
	0.10	0.11346	0.11346	0.11346	0.1135	0.1134
	0.05	0.02910	0.02910	0.02910	0.0291	0.0291
0.5	0.20	0.35530	0.35535	0.35533	0.3553	0.3606
	0.15	0.20913	0.20915	0.20914	0.2091	0.2121
	0.10	0.09636	0.09637	0.09637	0.0964	0.0975
	0.05	0.02466	0.02466	0.02466	0.0247	0.0249
1	0.20	0.32066	0.32070	0.32069	0.3207	0.3299
	0.15	0.18862	0.18864	0.18863	0.1886	0.1939
	0.10	0.08687	0.08687	0.08687	0.0869	0.0891
	0.05	0.02223	0.02223	0.02223	0.0222	0.0227
2	0.20	0.28935	0.28934	0.28934	0.2894	0.3016
	0.15	0.17067	0.17066	0.17066	0.1707	0.1778
	0.10	0.07879	0.07879	0.07879	0.0788	0.0819
	0.05	0.02020	0.02019	0.02019	0.0202	0.0209
5	0.20	0.26676	0.26657	0.26662	0.2668	0.2765
	0.15	0.15894	0.15887	0.15888	0.1589	0.1648
	0.10	0.07405	0.07403	0.07403	0.0740	0.0767
	0.05	0.01910	0.01910	0.01910	0.0191	0.0197

Table 3. Comparison of nondimensional frequency $\hat{\Omega}$ of FG plate resting on Winkler elastic foundation ($J_1 = 100$, $J_2 = 0$).

n	L/h	Present			Published	
		TDPT	SDPT	RDPT	Ref. [56]	Ref. [57]
0	0.20	0.42728	0.42735	0.42733	0.4273	0.4273
	0.15	0.25188	0.25190	0.25189	0.2519	0.2519
	0.10	0.11625	0.11626	0.11626	0.1163	0.1162
	0.05	0.02979	0.02979	0.02979	0.0298	0.0298
0.5	0.20	0.37048	0.37053	0.37051	0.3705	0.3758
	0.15	0.21744	0.21745	0.21745	0.2174	0.2204
	0.10	0.09998	0.09999	0.09999	0.1000	0.1012
	0.05	0.02556	0.02556	0.02556	0.0256	0.0258
1	0.20	0.33825	0.33829	0.33828	0.3382	0.3476
	0.15	0.19827	0.19829	0.19828	0.1983	0.2036
	0.10	0.09108	0.09109	0.09109	0.0911	0.0933
	0.05	0.02327	0.02327	0.02327	0.0233	0.0238
2	0.20	0.30977	0.30976	0.30976	0.3098	0.3219
	0.15	0.18186	0.18185	0.18185	0.1819	0.1889
	0.10	0.08367	0.08367	0.08367	0.0837	0.0867
	0.05	0.02140	0.02140	0.02140	0.0214	0.0221
5	0.20	0.29019	0.29002	0.29006	0.2902	0.2999
	0.15	0.17164	0.17157	0.17159	0.1716	0.1775
	0.10	0.07954	0.07952	0.07952	0.0795	0.0821
	0.05	0.02045	0.02045	0.02045	0.0205	0.0210

Table 4. Comparison of nondimensional frequency $\hat{\Omega}$ of FG plate resting on Pasternak elastic foundation ($J_1 = 100$, $J_2 = 100$).

n	L/h	Present			Published	
		TDPT	SDPT	RDPT	Ref. [56]	Ref. [57]
0	0.20	0.61618	0.61623	0.61621	0.6162	0.6162
	0.15	0.35604	0.35606	0.35605	0.3560	0.3560
	0.10	0.16188	0.16188	0.16188	0.1619	0.1619
	0.05	0.04110	0.04110	0.04110	0.0411	0.0411
0.5	0.20	0.59543	0.59545	0.59544	0.5954	0.6026
	0.15	0.34236	0.34237	0.34237	0.3424	0.3460
	0.10	0.15500	0.15500	0.15500	0.1550	0.1563
	0.05	0.03924	0.03924	0.03924	0.0392	0.0395
1	0.20	0.58553	0.58555	0.58554	0.5855	0.5978
	0.15	0.33613	0.33614	0.33613	0.3361	0.3422
	0.10	0.15197	0.15197	0.15197	0.1520	0.1542
	0.05	0.03844	0.03844	0.03844	0.0384	0.0388
2	0.20	0.58023	0.58023	0.58023	0.5802	0.5970
	0.15	0.33298	0.33298	0.33297	0.3330	0.3412
	0.10	0.15051	0.15051	0.15051	0.1505	0.1535
	0.05	0.03807	0.03807	0.03807	0.0381	0.0386
5	0.20	0.58349	0.58344	0.58345	0.5835	0.5993
	0.15	0.33495	0.33492	0.33493	0.3350	0.3427
	0.10	0.15153	0.15152	0.15152	0.1515	0.1543
	0.05	0.03836	0.03836	0.03836	0.0384	0.0388

Table 5. Comparison of nondimensional frequency ($\bar{\Omega} = L\omega\sqrt{\rho/C_{11}}$) of a homogeneous piezoelectromagnetic plate ($L = W = 60$ nm, $h = 5$ nm, $J_1 = J_2 = V_0 = P_0 = 0$).

m, l	Present	Published	
		Ref. [29]	Ref. [14]
1, 1	0.3830	0.3829	0.3698
1, 2	0.9330	0.9329	0.9247
2, 2	1.4571	1.4568	1.4800

6.2. Parametric Results

In this subsection, we will discuss the effects of the elastic foundation parameter, graphene weight fraction, power-law index, side-to-thickness ratio, external applied electric and magnetic potentials and piezoelectromagnetic multiple on the fundamental frequency of the nanocomposite piezoelectromagnetic rectangular plates under various boundary conditions.

Table 6 presented the effects of the boundary conditions on the nondimensional frequency Ω of the nanocomposite piezoelectromagnetic square plate for different values of the side-to-thickness ratio L/h and different plate types. Since the clamped condition enhances the plate strength, the frequency Ω increases as the clamped conditions increase.

Table 6. Nondimensional frequency Ω of a nanocomposite piezoelectromagnetic plate for different boundary conditions.

Type	L/h	SSSS	CSSS	CCSS	CCCS	CCCC
Type A	10	15.60604	20.17570	20.69866	24.26245	24.59618
	15	7.69551	9.79339	9.97522	11.66440	11.79149
	20	4.70668	5.90897	5.97521	6.96373	7.01009
	25	3.23406	4.01599	4.03570	4.68846	4.70050
	30	2.38941	2.94103	2.93993	3.40587	3.40216
Type B	10	13.91824	17.85522	17.87654	21.13514	20.96808
	15	6.64805	8.53231	8.55383	10.15398	10.10229
	20	3.94721	5.03870	5.04273	5.98381	5.95221
	25	2.64537	3.35515	3.34959	3.96789	3.94248
	30	1.91426	2.41261	2.40245	2.83992	2.81771
Type C	10	11.02065	14.12403	14.54444	16.95287	17.24037
	15	5.36317	6.79055	6.94550	8.08732	8.20204
	20	3.24892	4.06724	4.13150	4.79916	4.84784
	25	2.21630	2.74827	2.77484	3.21523	3.23516
	30	1.62823	2.00317	2.01206	2.32603	2.33213
Type D	10	10.54877	13.31021	13.62532	15.82989	16.05832
	15	5.16099	6.42669	6.52518	7.56204	7.63701
	20	3.14148	3.86961	3.90012	4.50712	4.53008
	25	2.15131	2.62703	2.63159	3.03336	3.03577
	30	1.58541	1.92249	1.91608	2.20366	2.19722

Figures 3 and 4 display the nondimensional frequency Ω of the nanocomposite piezoelectromagnetic square plate versus the side-to-thickness ratio L/h for various values of Winkler elastic foundation parameter J_1 and the shear elastic foundation parameter J_2. As expected, the increment of elastic foundation stiffness enhances the plate strength, leading to more frequency. It is also found that the impacts of the elastic foundation decrease as the plate becomes thinner.

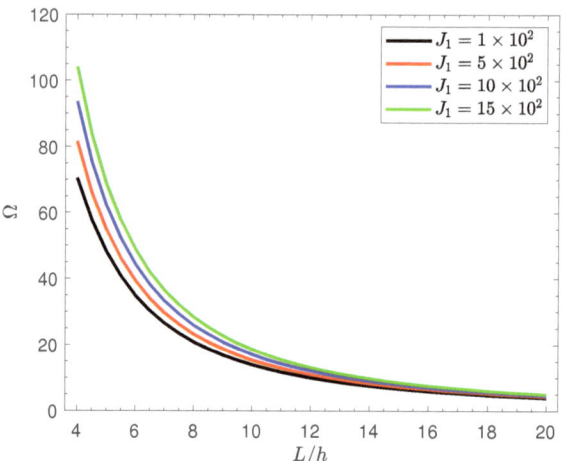

Figure 3. Effects of the Winkler elastic foundation parameter J_1 on the frequency of the nanocomposite piezoelectromagnetic plate (Type B).

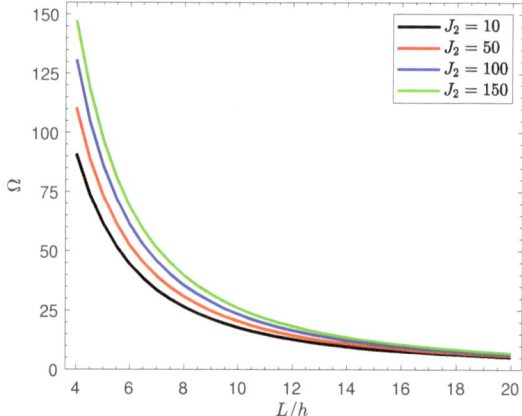

Figure 4. Effects of the shear elastic foundation parameter J_2 on the frequency of the nanocomposite piezoelectromagnetic plate (Type A).

Effects of the graphene weight fraction W_G on the fundamental frequency Ω of different types of the nanocomposite piezoelectromagnetic plate resting on an elastic foundation are revealed in Figure 5. It is evident from this figure that, regardless of the plate type, the frequency Ω monotonically decreases as the ratio L/h increases, because the plate becomes weaker with an increasing side-to-thickness ratio. Moreover, since the more flexible plate yields smaller frequencies, the decrease in the graphene components reduces the fundamental frequency Ω. The frequency of the types A, B, C and D can be arranged as: Type A > Type B > Type C > Type D. This means that the plate reinforced with uniformly distributed GNSs is stiffer than that reinforced with FG-GNSs. In addition, the influence of W_G is more pronounced in the case of Type A.

The variation in the fundamental frequency Ω of different types of the FG nanocomposite piezoelectromagnetic plate versus the plate aspect ratio W/L for different values of the power-law index k are displayed in Figure 6. It can be noticed that the frequency Ω gradually decreases as the aspect ratio W/L and the power-law index k increase.

Nondimensional frequencies Ω of the nanocomposite piezoelectromagnetic plate versus the side-to-thickness ratio L/h and the piezoelectromagnetic multiple χ for different values of the external applied electric and magnetic potentials (V_0 and P_0) are plotted in Figures 7 and 8. It can be observed that the applied electric potential V_0 has a softening effect on the nanocomposite piezoelectromagnetic plate, whereas the applied magnetic potential P_0 has a hardening effect. Accordingly, the frequency Ω is increased as the electric potential V_0 decreases and the magnetic potential P_0 increases. Further, it is seen that the frequency Ω predicted versus the piezoelectromagnetic multiple χ approach each other as the parameter χ decreases, which means that the electric and magnetic potentials lose their effects on the frequency.

For more explanation of the effects of the piezoelectromagnetic multiple χ on the fundamental frequency Ω, Figure 9 displays the frequency Ω against the side-to-thickness ratio L/h for various values of the parameter χ. It is notable that with the increase in the piezoelectromagnetic properties of graphene, the plate becomes stiffer, leading to an increment in the frequency.

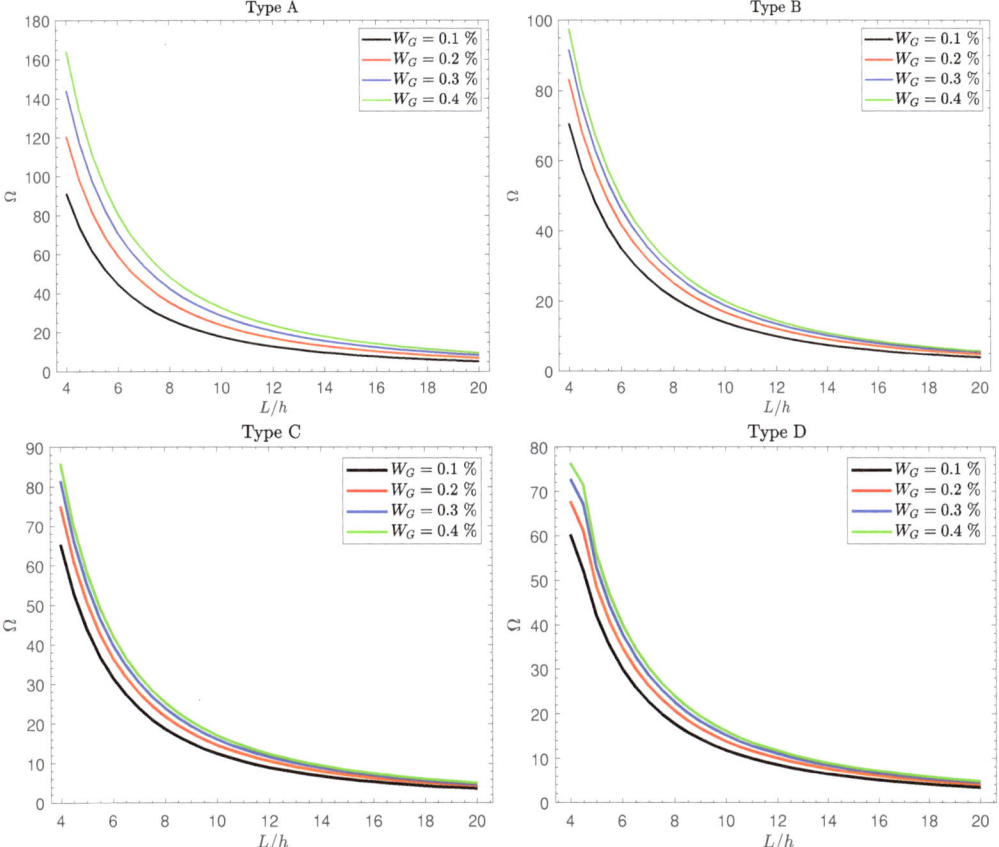

Figure 5. Effects of the graphene weight fraction W_G on the frequency of different types of the nanocomposite piezoelectromagnetic plate.

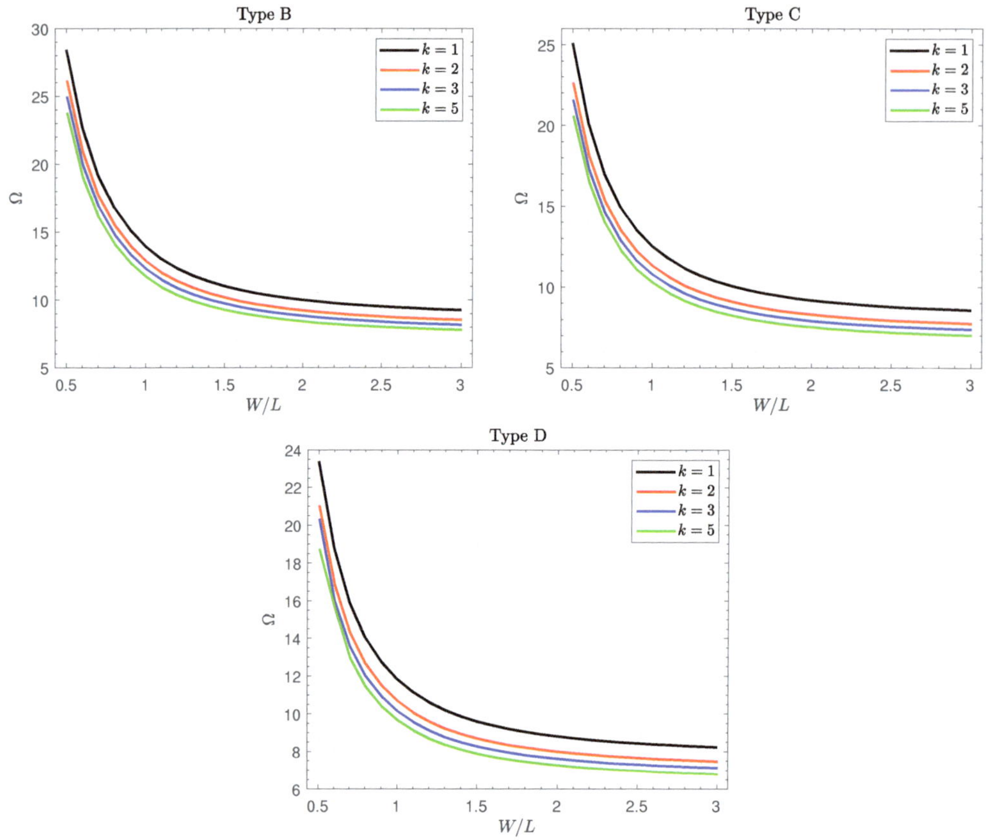

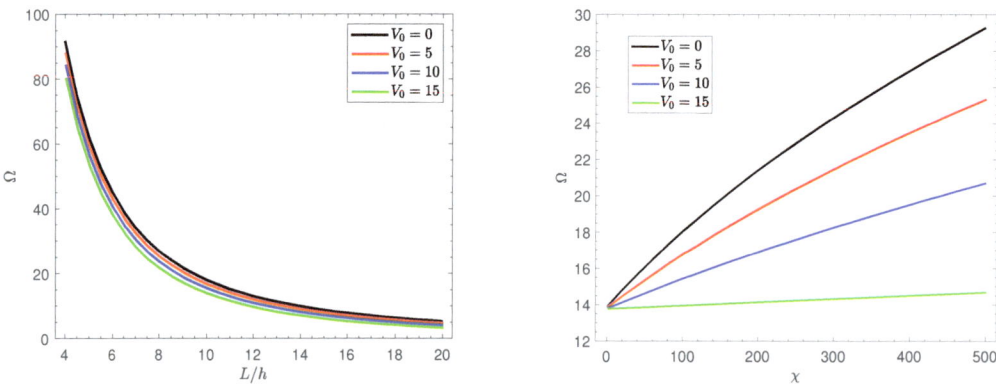

Figure 6. Effects of the power-law index k on the frequency of different types of the nanocomposite piezoelectromagnetic plate.

Figure 7. Frequency of the nanocomposite piezoelectromagnetic plate versus the side-to-thickness ratio L/h and the piezoelectromagnetic multiple χ for different values of the external applied electric potential V_0 (Type A).

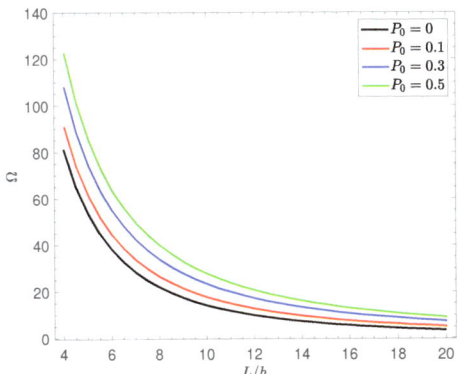
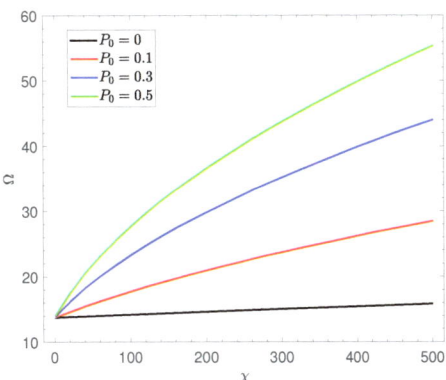

Figure 8. Frequency of the nanocomposite piezoelectromagnetic plate versus the side-to-thickness ratio L/h and the piezoelectromagnetic multiple χ for different values of the external applied magnetic potential P_0 (Type A).

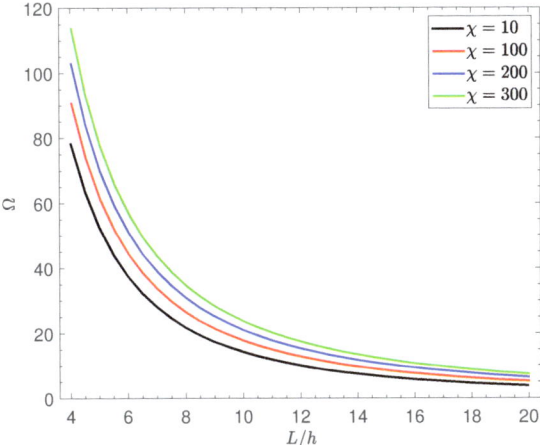

Figure 9. Effects of the piezoelectromagnetic multiple χ on the frequency of the nanocomposite piezoelectromagnetic plate (Type A).

7. Conclusions

Based on a refined four-unknown shear deformation plate theory, free vibration of piezoelectromagnetic plates reinforced with functionally graded graphene nanosheets (FG-GNSs) under simply supported conditions is analyzed. Winkler–Pasternak's elastic foundation is used as an elastic substrate of the present plate. The current nanocomposite plate is subjected to external applied electric and magnetic potentials. Hamilton's principle, including the electric displacements, the magnetic induction and elastic foundation reaction, is utilized to deduce the motion equations. Analytical solutions for simply-supported boundary conditions are employed based on a Navier-type solution. The current outcomes are compared with those published to develop the reliability of the present formulations. Moreover, various numerical examples are presented to illustrate the influences of several parameters on the fundamental frequency. It is found that increasing the elastic foundation stiffness, graphene weight fraction, applied magnetic potential and electromagnetic properties of graphene enhance the plate stiffness, leading to a noticeable increment in the fundamental frequency. Conversely, the increase in the side-to-thickness ratio, plate aspect ratio and applied electric potential weaken the plate strength, therefore the frequency decreases. In the same trend, the fundamental frequency is reduced with an increasing power-law index because the components of graphene decrease. In general, we can control the vibration of the present nanocomposite model via changing the different conditions, such as the applied electric potential, magnetic potential or the elastic foundation stiffness.

Author Contributions: Conceptualization, M.S. and F.H.H.A.M.; methodology, M.S. and F.H.H.A.M.; software, M.S.; validation, M.S. and F.H.H.A.M.; formal analysis, F.H.H.A.M.; investigation, F.H.H.A.M.; resources, F.H.H.A.M.; data curation, M.S; writing—original draft preparation, F.H.H.A.M.; writing—review and editing, M.S. and F.H.H.A.M.; visualization, F.H.H.A.M.; supervision, M.S. and F.H.H.A.M.; project administration, F.H.H.A.M.; funding acquisition, M.S. and F.H.H.A.M. All authors have read and agreed to the published version of the manuscript.

Funding: This work was supported through the Annual Funding track by the Deanship of Scientific Research, Vice Presidency for Graduate Studies and Scientific Research, King Faisal University, Saudi Arabia [Project No. AN00016].

Data Availability Statement: Not applicable.

Conflicts of Interest: The authors declare no conflict of interest.

References

1. Sun, J.; Xu, X.; Lim, C.W.; Zhou, Z.; Xiao, S. Accurate thermo-electro-mechanical buckling of shear deformable piezoelectric fiber-reinforced composite cylindrical shells. *Compos. Struct.* **2016**, *141*, 221–231. [CrossRef]
2. Sobhy, M. Piezoelectric bending of gpl-reinforced annular and circular sandwich nanoplates with fg porous core integrated with sensor and actuator using dqm. *Arch. Civ. Mech. Eng.* **2021**, *21*, 1–18. [CrossRef]
3. Phung-Van, P.; Tran, L.V.; Ferreira, A.J.M.; Nguyen-Xuan, H.; Abdel-Wahab, M.J.N.D. Nonlinear transient isogeometric analysis of smart piezoelectric functionally graded material plates based on generalized shear deformation theory under thermo-electro-mechanical loads. *Nonlinear Dyn.* **2017**, *87*, 879–894. [CrossRef]
4. Phung-Van, P.; Nguyen, L.B.; Tran, L.V.; Dinh, T.D.; Thai, C.H.; Bordas, S.P.A.; Abdel-Wahab, M.; Nguyen-Xuan, H. An efficient computational approach for control of nonlinear transient responses of smart piezoelectric composite plates. *Int. J. Non-Linear Mech.* **2015**, *76*, 190–202. [CrossRef]
5. Phung-Van, P.; De Lorenzis, L.; Thai, C.H.; Abdel-Wahab, M.; Nguyen-Xuan, H. Analysis of laminated composite plates integrated with piezoelectric sensors and actuators using higher-order shear deformation theory and isogeometric finite elements. *Comput. Mater. Sci.* **2015**, *96*, 495–505. [CrossRef]
6. Sahu, M.; Hajra, S.; Lee, K.; Deepti, P.L.; Mistewicz, K.; Kim, H.J. Piezoelectric nanogenerator based on lead-free flexible PVDF-barium titanate composite films for driving low power electronics. *Crystals* **2021**, *11*, 85. [CrossRef]
7. Yan, X.; Zheng, M.; Zhu, M.; Hou, Y. Soft and hard piezoelectric ceramics for vibration energy harvesting. *Crystals* **2020**, *10*, 907. [CrossRef]
8. Hu, K.-Q.; Li, G.-Q. Electro-magneto-elastic analysis of a piezoelectromagnetic strip with a finite crack under longitudinal shear. *Mech. Mater.* **2005**, *37*, 925–934. [CrossRef]
9. Liu, H.; Zhang, Q.; Yang, X.; Ma, J. Size-dependent vibration of laminated composite nanoplate with piezo-magnetic face sheets. *Eng. Comput.* **2021**, 1–17. [CrossRef]
10. Bin, W.; Jiangong, Y.; Cunfu, H. Wave propagation in non-homogeneous magneto-electro-elastic plates. *J. Sound Vib.* **2008**, *317*, 250–264. [CrossRef]
11. Khorasani, M.; Soleimani-Javid, Z.; Arshid, E.; Amir, S.; Civalek, O. Vibration analysis of graphene nanoplatelets' reinforced composite plates integrated by piezo-electromagnetic patches on the piezo-electromagnetic media. *Waves Random Complex Media* **2021**, 1–31. [CrossRef]
12. Naskar, S.; Shingare, K.B.; Mondal, S.; Mukhopadhyay, T. Flexoelectricity and surface effects on coupled electromechanical responses of graphene reinforced functionally graded nanocomposites: A unified size-dependent semi-analytical framework. *Mech. Syst. Signal Process.* **2022**, *169*, 108757. [CrossRef]
13. Samadi, A.; Hosseini, S.M.; Mohseni, M. Investigation of the electromagnetic microwaves absorption and piezoelectric properties of electrospun Fe_3O_4-GO/PVDF hybrid nanocomposites. *Org. Electron.* **2018**, *59*, 149–155. [CrossRef]
14. Ke, L.-L.; Wang, Y.-S.; Yang, J.; Kitipornchai, S. Free vibration of size-dependent magneto-electro-elastic nanoplates based on the nonlocal theory. *Acta Mech. Sin.* **2014**, *30*, 516–525. [CrossRef]
15. Ke, L.-L.; Wang, Y.-S. Free vibration of size-dependent magneto-electro-elastic nanobeams based on the nonlocal theory. *Phys. E Low-Dimens. Syst. Nanostruct.* **2014**, *63*, 52–61. [CrossRef]
16. Li, Y.S.; Ma, P.; Wang, W. Bending, buckling, and free vibration of magnetoelectroelastic nanobeam based on nonlocal theory. *J. Intell. Mater. Syst. Struct.* **2016**, *27*, 1139–1149. [CrossRef]
17. Pan, E.; Han, F. Exact solution for functionally graded and layered magneto-electro-elastic plates. *Int. J. Eng. Sci.* **2005**, *43*, 321–339. [CrossRef]

18. Farajpour, A.; Yazdi, M.R.H.; Rastgoo, A.; Loghmani, M.; Mohammadi, M. Nonlocal nonlinear plate model for large amplitude vibration of magneto-electro-elastic nanoplates. *Compos. Struct.* **2016**, *140*, 323–336. [CrossRef]
19. Farajpour, A.; Rastgoo, A.; Farajpour, M.R. Nonlinear buckling analysis of magneto-electro-elastic cnt-mt hybrid nanoshells based on the nonlocal continuum mechanics. *Compos. Struct.* **2017**, *180*, 179–191. [CrossRef]
20. Jamalpoor, A.; Ahmadi-Savadkoohi, A.; Hosseini, M.; Hosseini-Hashemi, S. Free vibration and biaxial buckling analysis of double magneto-electro-elastic nanoplate-systems coupled by a visco-pasternak medium via nonlocal elasticity theory. *Eur. J. Mech. A/Solids* **2017**, *63*, 84–98. [CrossRef]
21. Mehditabar, A.; Rahimi, G.H.; Sadrabadi, S.A. Three-dimensional magneto-thermo-elastic analysis of functionally graded cylindrical shell. *Appl. Math. Mech.* **2017**, *38*, 479–494. [CrossRef]
22. Zenkour, A.M.; Aljadani, M.H. Buckling analysis of actuated functionally graded piezoelectric plates via a quasi-3d refined theory. *Mech. Mater.* **2020**, *151*, 103632. [CrossRef]
23. Meskini, M.; Ghasemi, A.R. Electro-magnetic potential effects on free vibration of rotating circular cylindrical shells of functionally graded materials with laminated composite core and piezo electro-magnetic two face sheets. *J. Sandw. Struct. Mater.* **2021**, *23*, 2772–2797. [CrossRef]
24. Abazid, M.A.; Sobhy, M. Thermo-electro-mechanical bending of FG piezoelectric microplates on Pasternak foundation based on a four-variable plate model and the modified couple stress theory. *Microsyst. Technol.* **2018**, *24*, 1227–1245. [CrossRef]
25. Monaco, G.T.; Fantuzzi, N.; Fabbrocino, F.; Luciano, R. Critical temperatures for vibrations and buckling of magneto-electro-elastic nonlocal strain gradient plates. *Nanomaterials* **2021**, *11*, 87. [CrossRef]
26. Monaco, G.T.; Fantuzzi, N.; Fabbrocino, F.; Luciano, R. Trigonometric solution for the bending analysis of magneto-electro-elastic strain gradient nonlocal nanoplates in hygro-thermal environment. *Mathematics* **2021**, *9*, 567. [CrossRef]
27. Chen, J.; Guo, J.; Pan, E. Wave propagation in magneto-electro-elastic multilayered plates with nonlocal effect. *J. Sound Vib.* **2017**, *400*, 550–563. [CrossRef]
28. Ebrahimi, F.; Dabbagh, A. Wave dispersion characteristics of rotating heterogeneous magneto-electro-elastic nanobeams based on nonlocal strain gradient elasticity theory. *J. Electromagn. Waves Appl.* **2018**, *32*, 138–169. [CrossRef]
29. Abazid, M.A. The nonlocal strain gradient theory for hygrothermo-electromagnetic effects on buckling, vibration and wave propagation in piezoelectromagnetic nanoplates. *Int. J. Appl. Mech.* **2019**, *11*, 1950067. [CrossRef]
30. Arefi, M.; Zenkour, A.M. Wave propagation analysis of a functionally graded magneto-electro-elastic nanobeam rest on visco-pasternak foundation. *Mech. Res. Commun.* **2017**, *79*, 51–62. [CrossRef]
31. Sobhy, M. Analytical buckling temperature prediction of fg piezoelectric sandwich plates with lightweight core. *Mater. Res. Express* **2021**, *8*, 095704. [CrossRef]
32. Sobhy, M. Stability analysis of smart FG sandwich plates with auxetic core. *Int. J. Appl. Mech.* **2021**, *13*, 2150093. [CrossRef]
33. Potts, J.R.; Dreyer, D.R.; Bielawski, C.W.; Ruoff, R.S. Graphene-based polymer nanocomposites. *Polymer* **2011**, *52*, 5–25. [CrossRef]
34. Papageorgiou, D.G.; Kinloch, I.A.; Young, R.J. Mechanical properties of graphene and graphene-based nanocomposites. *Prog. Mater. Sci.* **2017**, *90*, 75–127. [CrossRef]
35. Yang, J.; Zhang, Y.; Li, Y.; Wang, Z.; Wang, W.; An, Q.; Tong, W. Piezoelectric nanogenerators based on graphene oxide/pvdf electrospun nanofiber with enhanced performances by in-situ reduction. *Mater. Today Commun.* **2021**, *26*, 101629. [CrossRef]
36. Forsat, M.; Musharavati, F.; Eltai, E.; Zain, A.M.; Mobayen, S.; Mohamed, A.M. Vibration characteristics of microplates with GNPs-reinforced epoxy core bonded to piezoelectric-reinforced CNTs patches. *Adv. Nano Res.* **2021**, *11*, 115–140.
37. Thai, C.H.; Phung-Van, P. A meshfree approach using naturally stabilized nodal integration for multilayer FG GPLRC complicated plate structures. *Eng. Anal. Bound. Elem.* **2020**, *117*, 346–358. [CrossRef]
38. Thai, C.H.; Ferreira, A.J.M.; Tran, T.D.; Phung-Van, P. Free vibration, buckling and bending analyses of multilayer functionally graded graphene nanoplatelets reinforced composite plates using the NURBS formulation. *Compos. Struct.* **2019**, *220*, 749–759. [CrossRef]
39. Phung-Van, P.; Lieu, Q.X.; Ferreira, A.J.M.; Thai, C.H. A refined nonlocal isogeometric model for multilayer functionally graded graphene platelet-reinforced composite nanoplates. *Thin-Walled Struct.* **2021**, *164*, 107862. [CrossRef]
40. Mao, J.J.; Lu, H.M.; Zhang, W.; Lai, S.K. Vibrations of graphene nanoplatelet reinforced functionally gradient piezoelectric composite microplate based on nonlocal theory. *Compos. Struct.* **2020**, *236*, 111813. [CrossRef]
41. Sobhy, M. Magneto-electro-thermal bending of fg-graphene reinforced polymer doubly-curved shallow shells with piezoelectromagnetic faces. *Compos. Struct.* **2018**, *203*, 844–860. [CrossRef]
42. Mao, J.-J.; Zhang, W. Linear and nonlinear free and forced vibrations of graphene reinforced piezoelectric composite plate under external voltage excitation. *Compos. Struct.* **2018**, *203*, 551–565. [CrossRef]
43. Mao, J.-J.; Zhang, W. Buckling and post-buckling analyses of functionally graded graphene reinforced piezoelectric plate subjected to electric potential and axial forces. *Compos. Struct.* **2019**, *216*, 392–405. [CrossRef]
44. Sobhy, M.; Abazid, M.A.; Mukahal, F.H.H.A. Electro-thermal buckling of fg graphene platelets-strengthened piezoelectric beams under humid conditions. *Adv. Mech. Eng.* **2022**. [CrossRef]
45. Abolhasani, M.M.; Shirvanimoghaddam, K.; Naebe, M. Pvdf/graphene composite nanofibers with enhanced piezoelectric performance for development of robust nanogenerators. *Compos. Sci. Technol.* **2017**, *138*, 49–56. [CrossRef]
46. Xu, K.; Wang, K.; Zhao, W.; Bao, W.; Liu, E.; Ren, Y.; Wang, M.; Fu, Y.; Zeng, J.; Li, Z.; et al. The positive piezoconductive effect in graphene. *Nat. Commun.* **2015**, *6*, 1–6. [CrossRef]

47. Al Mukahal, F.H.H.; Sobhy, M. Wave propagation and free vibration of FG graphene platelets sandwich curved beam with auxetic core resting on viscoelastic foundation via DQM. *Arch. Civ. Mech. Eng.* **2022**, *22*, 1–21. [CrossRef]
48. Allam, M.N.M.; Radwan, A.F.; Sobhy, M. Hygrothermal deformation of spinning FG graphene sandwich cylindrical shells having an auxetic core. *Eng. Struct.* **2022**, *251*, 113433. [CrossRef]
49. Sobhy, M.; Alakel Abazid, M. Mechanical and thermal buckling of FG-GPLs sandwich plates with negative Poisson's ratio honeycomb core on an elastic substrate. *Eur. Phys. J. Plus* **2022**, *137*, 1–21. [CrossRef]
50. Shimpi, R.P. Refined plate theory and its variants. *AIAA J.* **2002**, *40*, 137–146. [CrossRef]
51. Zenkour, A.M.; Sobhy, M. Axial magnetic field effect on wave propagation in bi-layer fg graphene platelet-reinforced nanobeams. *Eng. Comput.* **2021**, *37*, 1–17. [CrossRef]
52. Reddy, J.N. A simple higher-order theory for laminated composite plates. *J. Appl. Mech. Dec.* **1984**, *51*, 745–752. [CrossRef]
53. Touratier, M. An efficient standard plate theory. *Int. J. Eng. Sci.* **1991**, *29*, 901–916. [CrossRef]
54. Karama, M.; Afaq, K.S.; Mistou, S. Mechanical behaviour of laminated composite beam by the new multi-layered laminated composite structures model with transverse shear stress continuity. *Int. J. Solids Struct.* **2003**, *40*, 1525–1546. [CrossRef]
55. Zhang, S.; Xia, R.; Lebrun, L.; Anderson, D.; Shrout, T.R. Piezoelectric materials for high power, high temperature applications. *Mater. Lett.* **2005**, *59*, 3471–3475. [CrossRef]
56. Thai, H.-T.; Choi, D.-H. A refined plate theory for functionally graded plates resting on elastic foundation. *Compos. Sci. Technol.* **2011**, *71*, 1850–1858. [CrossRef]
57. Hasani Baferani, A; Saidi, A.R.; Ehteshami, H. Accurate solution for free vibration analysis of functionally graded thick rectangular plates resting on elastic foundation. *Compos. Struct.* **2011**, *93*, 1842–1853. [CrossRef]

MDPI AG
Grosspeteranlage 5
4052 Basel
Switzerland
Tel.: +41 61 683 77 34

Crystals Editorial Office
E-mail: crystals@mdpi.com
www.mdpi.com/journal/crystals

Disclaimer/Publisher's Note: The title and front matter of this reprint are at the discretion of the Guest Editors. The publisher is not responsible for their content or any associated concerns. The statements, opinions and data contained in all individual articles are solely those of the individual Editors and contributors and not of MDPI. MDPI disclaims responsibility for any injury to people or property resulting from any ideas, methods, instructions or products referred to in the content.

www.ingramcontent.com/pod-product-compliance
Lightning Source LLC
LaVergne TN
LVHW070002100526
838202LV00019B/2607